Praise for *The Compass of Pleasure*

"In his book *The Compass of Pleasure*, the Johns Hopkins neurobiologist David J. Linden explicates the workings of [the regions of the brain] known collectively as the reward system, elegantly drawing on sources ranging from personal experience to studies of brain activity to experiments with molecules and genes." —*The New York Times Book Review*

"Hugely entertaining . . . If you're science-phobic, don't worry: Linden is incredibly smart, but comes across as the funny, patient professor you wish you'd had in college." —NPR.org

"How do orgasms, heroin, greasy foods, and juicy gossip jolt the same neurons? Neuroscientist David Linden delves into the research, mixing in plenty of trippy anecdotes." —*Psychology Today*

"Linden's conversational style, his abundant use of anecdotes, and his successful coupling of wit with insight makes the book a joy to read. Even the footnotes are sprinkled with hidden gems." —*Publishers Weekly*

"Conventional wisdom advises, 'If it feels good, stop it. If it tastes good, spit it out.' But why? Because indulging pleasurable excess, whether of drugs, food, or sex, has an unforgiving downside. The biology of how we know this is the topic of Linden's fascinating, by turns technical and entertaining, effort." —Donna Chavez, *Booklist*

"This cheerful summary of the brain's reward system is a profound experience. . . . *Pleasure* is a superb book. My brain has been changed by reading it." —Leo Benedictus, *The Guardian* (London)

"This book is highly readable and full of fascinating facts and theories. . . . You're sure to get pleasure from reading *Pleasure*." —Susan Blackmore, *BBC Focus* (London)

ABOUT THE AUTHOR

David J. Linden is a professor of neuroscience at The Johns Hopkins University School of Medicine. The author of *The Accidental Mind* (2007), *The Compass of Pleasure* (2011), and most recently, *Touch* (2015), he served for many years as the chief editor of *The Journal of Neurophysiology*. He lives in Baltimore, Maryland, with his two children.

THE
COMPASS
OF
PLEASURE

How Our Brains Make Fatty Foods, Orgasm,

Exercise, Marijuana, Generosity, Vodka,

Learning, and Gambling

Feel So Good

DAVID J. LINDEN

PENGUIN BOOKS

PENGUIN BOOKS

Published by the Penguin Group

Penguin Group (USA) Inc., 375 Hudson Street, New York, New York 10014, U.S.A.
Penguin Group (Canada), 90 Eglinton Avenue East, Suite 700, Toronto,
Ontario, Canada M4P 2Y3 (a division of Pearson Penguin Canada Inc.)
Penguin Books Ltd, 80 Strand, London WC2R 0RL, England
Penguin Ireland, 25 St Stephen's Green, Dublin 2, Ireland (a division of Penguin Books Ltd)
Penguin Books Australia Ltd, 250 Camberwell Road, Camberwell, Victoria 3124, Australia
(a division of Pearson Australia Group Pty Ltd)
Penguin Books India Pvt Ltd, 11 Community Centre,
Panchsheel Park, New Delhi – 110 017, India
Penguin Group (NZ), 67 Apollo Drive, Rosedale, Auckland 0632,
New Zealand (a division of Pearson New Zealand Ltd)
Penguin Books (South Africa) (Pty) Ltd, 24 Sturdee Avenue,
Rosebank, Johannesburg 2196, South Africa

Penguin Books Ltd, Registered Offices: 80 Strand, London WC2R 0RL, England

First published in the United States of America by Viking Penguin,
a member of Penguin Group (USA) Inc. 2011
Published in Penguin Books 2012

7   9   10   8   6

Excerpt from *Knockemstiff* by Donald Ray Pollock. Copyright © 2008 by Donald Ray
Pollock. Used by permission of Doubleday, a division of Random House, Inc.

THE LIBRARY OF CONGRESS HAS CATALOGED THE HARDCOVER EDITION AS FOLLOWS:
Linden, David J.
The compass of pleasure : how our brains make fatty foods, orgasm, exercise, marijuana,
generosity, vodka, learning, and gambling feel so good / David J. Linden.
p. cm.
Includes bibliographical references and index.
ISBN 978-0-670-02258-8 (hc.)
ISBN 978-0-14-312075-9 (pbk.)
1. Pleasure—Physiological aspects.   2. Neuropsychology.   I. Title.
QP401.L56 2011
612.8—dc22   2010035380

Printed in the United States of America

For Elaine Levin

CONTENTS

"Pleasure never comes sincere to man;
but lent by heaven upon hard usury."

—John Dryden, *Edippus*

"Phil was probably passed out somewhere, enjoying his
dead father's legacy. I found myself wishing I had a
loved one who would die and leave me their barbitu-
rates, but I couldn't think of anyone who'd ever loved
me that much. My uncle had already promised his to the
mail lady."

—Donald Ray Pollock, "Bactine"

PROLOGUE

Bangkok, 1989. The afternoon rains have ended, leaving the early evening air briefly free of smog and allowing that distinctive Thai perfume, frangipani with a faint note of sewage, to waft over the shiny streets. I hail a *tuk-tuk*, a three-wheel motorcycle taxi, and hop aboard. My young driver has an entrepreneurial smile as he turns around and begins the usual interrogation of male travelers.

"So . . . you want girl?"

"No."

"I see." Long pause, eyebrows slowly raised. "You want boy!"

"Uh, no."

Longer pause. Sound of engine sputtering at idle. "You want ladyboy?"

"No," I answer, a bit more emphatically, nonplussed at the idea that I give the impression of desiring this particular commodity.

"I got cheap cigarettes . . . Johnnie Walker . . ."

"No thanks."

Undaunted, he moves on to the next category of his wares, now with lowered voice.

"You want ganja?"

"No."

"Coke?"

"No."

"*Ya baa* [methamphetamine tablets]?"

"Nope."

A whisper now. "Heroin?"

"No."

Voice raised back to normal. "I can take you to cockfight. You can gamble!"

"I'll pass."

Just a little bit irritated now. "So, *farang*, what you want?"

"*Prik kee noo*," I respond. "Those little 'mouse shit' peppers. I want some good, spicy dinner." My driver, not surprisingly, is disappointed. As we tear through the streets to a restaurant, blasting through puddles, I'm left wondering: *Aside from various shades of illegality, what do all his offers have in common? What is it exactly that makes a vice?*

~~~~~

We humans have a complicated and ambivalent relationship to pleasure, which we spend an enormous amount of time and resources pursuing. A key motivator of our lives, pleasure is central to learning, for we must find things like food, water, and sex rewarding in order to survive and pass our genetic material to the next generation. Certain forms of pleasure are accorded special status. Many of our most important rituals involving prayer, music, dance, and meditation produce a kind of transcendent pleasure that has become deeply ingrained in human cultural practice.

As we do with most powerful forces, however, we also want to regulate pleasure. In cultures around the world we find well-defined ideas and rules about pleasure that have persisted throughout history in any number of forms and variations:

Pleasure should be sought in moderation.
Pleasure must be earned.
Pleasure must be achieved naturally.
Pleasure is transitory.
The denial of pleasure can yield spiritual growth.

Our legal systems, our religions, our educational systems are all deeply concerned with controlling pleasure. We have created detailed rules and customs surrounding sex, drugs, food, alcohol, and even gambling. Jails are bursting with people who have violated laws that proscribe certain forms of pleasure or who profit by encouraging others to do so.

One can fashion reasonable theories of human pleasure and its regulation using the methods of cultural anthropology or social history. These are valid and useful endeavors, for ideas and practices involving human pleasure are certainly deeply influenced by culture. However, what I'm seeking here in *The Compass of Pleasure* is a different type of understanding—one less nuanced, perhaps, but more fundamental: a cross-cultural biological explanation. In this book I will argue that most experiences in our lives that we find transcendent—whether illicit vices or socially sanctioned ritual and social practices as diverse as exercise, meditative prayer, or even charitable giving—activate an anatomically and biochemically defined pleasure circuit in the brain. Shopping, orgasm, learning, highly caloric foods, gambling, prayer, dancing 'til you drop, and playing on the Internet: They all evoke neural signals that converge on a small group of interconnected brain areas called the medial forebrain pleasure circuit. It is in these tiny clumps of neurons that human pleasure is felt. This intrinsic pleasure circuitry can also be co-opted by artificial activators like cocaine or nicotine or heroin or alcohol. Evolution has, in effect, hardwired us to catch a pleasure buzz from a wide variety of experiences from crack to cannabis, from meditation to masturbation, from Bordeaux to beef.

This theory of pleasure reframes our understanding of the part of the human body that societies are most intent upon regulating. While we might assume that the anatomical region most closely governed by laws, religious prohibitions, and social mores is the genitalia, or the mouth, or the vocal cords, it is actually the medial forebrain pleasure circuit. As societies and as individuals, we are hell-bent on achieving and controlling pleasure, and it is those neurons, deep in our brains, that are the nexus of that struggle.

These particular neurons also comprise another battleground. The dark side of pleasure is, of course, addiction. It is now becoming clear that addiction is associated with long-lasting changes in the electrical, morphological, and biochemical functions of neurons and synaptic connections within the medial forebrain pleasure circuit. There are strong suggestions that these changes underlie many of the terrifying aspects of addiction, including tolerance (needing successively larger doses to get high), craving, withdrawal, and relapse. Provocatively, such persistent changes appear to be nearly identical to experience- and learning-driven changes in neural circuitry that are used to store memories in other brain regions. In this way, memory, pleasure, and addiction are closely intertwined.

However, addiction is not the only force responsible for experience-driven changes within the brain's pleasure circuits. The combination of associative learning and pleasure has created nothing less than a cognitive miracle: We can be motivated by pleasure to achieve goals that are entirely arbitrary—goals that may or may not have an evolutionary adaptive value. These can be as wide-ranging as reality-based television and curling. For us humans (and probably for other primates and for cetaceans as well), even mere *ideas* can activate the pleasure circuit. Our eclecticism where pleasure is concerned serves to make our human existence wonderfully rich and complex.

~~~~~

I like to tell the students in my lab that the golden age of brain research is right now, so it's time to get down to business. This sounds like a cheap motivational gimmick, but it's true. Our accumulating understanding of neural function, coupled with enabling technologies that allow us to measure and manipulate the brain with unprecedented precision, has given us new and often counterintuitive insights into behavioral and cognitive phenomena at the levels of biological processes. Nowhere is this more evident than in the neurobiology of pleasure. One example: Do you, like many, think that drug addicts become drug addicts because they derive greater reward from getting high than others? The biology says no: They actually seem to *want* it more but *like* it less.

This level of analysis is not only of academic interest. Understanding the biological basis of pleasure leads us to fundamentally rethink the moral and legal aspects of addiction to drugs, food, sex, and gambling and the industries that manipulate these pleasures in the marketplace. It also calls for a reformation in our concepts of such virtuous and prosocial behaviors as sharing resources, self-deprivation, and the drive for knowledge. Crucially, brain imaging studies show that giving to charity, paying taxes, and receiving information about future events all activate the same neural pleasure circuit that's engaged by heroin or orgasm or fatty foods. Perhaps, most important, analysis of the molecular basis of enduring changes in the brain's pleasure circuitry holds great promise for developing drugs and other therapies to help people break free of addictions of many sorts, to both substances and experiences.

When I was a postdoctoral fellow at the Roche Institute of Molecular Biology in the early 1990s, I was fortunate to work with Sid Udenfriend, a pioneer in the biochemistry of the brain and a real mensch. Sid's favorite pedagogical phrase, usually intoned at the bar, was "It's always good to know a little chemistry." I couldn't agree more. It would be possible to write a book exploring the brain's pleasure circuits that was free of not only molecules but also basic anatomy, but that sort of spoon-feeding would require ignoring some of the most interesting and important issues, and so that's not what you'll find here. If you come along for the ride and work with me just a bit to learn some basic neuroscience, I'll do my best to make it lively and fun as we explore the cellular and molecular basis of human pleasure, transcendent experience, and addiction.

CHAPTER ONE

MASHING THE PLEASURE BUTTON

ontréal, 1953. Fortunately, Peter Milner and James Olds didn't have perfect aim. While postdoctoral fellows at McGill University, under the direction of the renowned psychologist Donald Hebb, Olds and Milner were conducting experiments that involved implanting electrodes deep in the brains of rats. The implanting surgery, conducted while the animals were anesthetized, involved cementing a pair of electrodes half a millimeter apart to their skulls. After a few days of recovery from the surgery, the rats were fine. Long, flexible wires were then attached to the electrodes at one end and to an electrical stimulator at the other, to allow for activation of the specific brain region where the tips of the electrodes had come to rest.

One fall day Olds and Milner were testing a rat in which they had attempted to target a structure called the midbrain reticular system. Located at the midline of the brain, at the point where its base tapers to form the brain stem, this region had previously been shown by another lab to control sleeping and waking cycles. In this particular surgery, however, the electrodes had gone astray and come to rest still at the midline, but at a somewhat more forward position in the brain, in a region called the septum.

The rat in question was placed in a large rectangular box with corners labeled A, B, C, and D and was allowed to explore freely. Whenever the rat went to corner A, Olds pressed a button that delivered a brief, mild electrical shock through the implanted electrodes. (Unlike the rest of the body, brain tissue does not have the receptors that allow for pain detection, so such shocks don't produce a painful sensation within the skull.) After a few jolts, the rat kept returning to corner A and finally fell asleep in a different location. The next day, however, the rat seemed even more interested in corner A than the others. Olds and Milner were excited: They believed that they had found a brain region that, when stimulated, provoked general curiosity. However, further experiments on this same rat soon proved that not to be the case. By this time, the rat had acquired a habit of returning often to corner A to be stimulated. The researchers then tried to coax the rat away from corner A by administering a shock every time the rat made a step in the direction of corner B. This worked all too well—within five minutes, the rat relocated to corner B. Further investigation revealed that this rat could be directed to any location within the box with well-timed brain shocks—brief ones to guide the rat to the target location and then more sustained ones once it arrived there.

Many years earlier the psychologist B. F. Skinner had devised the operant conditioning chamber, or "Skinner box," in which a lever press by an animal triggered either a reinforcing stimulus, such as delivery of food or water, or a punishing stimulus, such as a painful foot shock. Rats placed in a Skinner box will rapidly learn to press a lever for a food reward and to avoid pressing a lever that delivers the shock. Olds and Milner now modified the chamber so that a lever press would deliver direct brain stimulation through the implanted electrodes. What resulted was perhaps the most dramatic experiment in the history of behavioral neuroscience: Rats would press the lever as many as seven thousand times per hour to stimulate their brains. They weren't stimulating a "curiosity center" at all—this was a pleasure center, a reward circuit, the activation of which was much more powerful than any natural stimulus. A series of subsequent experiments revealed that rats preferred pleasure circuit stimulation to food (even when they were hungry) and water (even when they were thirsty). Self-stimulating male rats would ignore a female in heat and would repeatedly cross foot-shock-delivering floor grids to reach the lever. Female rats would abandon their newborn nursing pups to continually press the lever. Some rats would self-stimulate as often as two thousand times per hour for twenty-four hours, to the exclusion of all other activities. They had to be unhooked from the apparatus to prevent death by self-starvation. Pressing that lever became their entire world (Figure 1.1).

Further work was done to systematically vary the placement of the electrode tips and thereby map the reward circuits of the brain. These experiments revealed that stimulation of the outer (and upper) surface of the brain, the neocortex, where sensory and motor processing mostly reside, produced no reward—the rats continued to press the lever at chance levels. However, deep in the brain, there was not just a single discrete location underlying reward. Rather, a group of interconnected structures, all located near the base of the brain and distributed along the midline, constituted the reward circuit. These included the ventral tegmental area, the nucleus accumbens, the medial forebrain bundle, and the septum, as well as portions of the thalamus and hypothalamus (more on these various regions later). Not all these areas were equally rewarding. Stimulation in some parts of this medial forebrain pleasure circuit could support self-stimulation rates of seven thousand lever presses per hour, while others elicited only two hundred per hour.

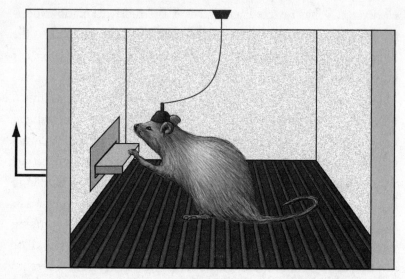

Figure 1.1 Self-stimulation of the pleasure circuit in a rat. When the rat presses the lever, it causes brief electrical stimulation to travel down the wire and activate the electrodes implanted deep in the rat's brain, in various portions of the medial forebrain pleasure circuit. This setup can be modified in several useful ways. For example, the electronics can be configured so that a rat must make many lever presses to get a single stimulation. In addition, a hollow needle can be implanted together with the stimulating electrodes to inject drugs directly into the pleasure circuit. Illustration by Joan M. K. Tycko.

It's hard to imagine now, but in 1953 the notion that motivational or pleasure/reward mechanisms could be localized to certain brain regions or circuits was highly controversial. The dominant theory, which had held sway for many years, was that excitation of the brain was always punishing and that learning and the development of behavior could be explained solely by punishment avoidance. This was called the drive-reduction hypothesis. In Olds's characterization of the theory, "pain supplies the push and learning based on pain reduction supplies the direction." There was no need for reward or pleasure: This model was all stick, no carrot. The pioneering experiments of Olds and Milner clearly demolished the punishment-only model in favor of a more comprehensive, hedonistic view that "behavior is pulled forward by pleasure as well as pushed forward by pain."[1]

~~~~~

I know what you're thinking: What does it feel like for a human to have his or her medial forebrain reward circuitry stimulated with an electrode? Does it produce a crazy pleasure that's better than food or sex or sleep or even *Seinfeld* reruns? We do in fact know the answer to that question. The bad news is that that answer comes, in part, from some deeply unethical experiments.

Dr. Robert Galbraith Heath was the founder and chairman of the Department of Psychiatry and Neurology at Tulane University in New Orleans. He served from 1949 to 1980, and during that time the major focus of his work involved stimulation of the brains of institutionalized psychiatric patients, often African Americans, using surgically implanted electrodes. His main goal—to use brain stimulation to relieve the symptoms of psychiatric disorders such as major depression and schizophrenia—was laudable. However, he did not obtain proper informed consent from his patients and took decisions in experimental design that would never be approved by modern human-subjects ethical review boards.

Perhaps the most egregious example was reported in a paper entitled "Septal stimulation for the initiation of heterosexual behavior in a homosexual male," published in the *Journal of Behavioral Therapy and Experimental Psychiatry* in 1972.[2] The rationale behind this experiment was that because stimulation of the septal area evoked pleasure, if it was combined with heterosexual imagery it could "bring about heterosexual behavior in a fixed, overt homosexual male." And so Patient B-19, a twenty-four-year-old male homosexual of average intelligence who suffered from depression and obsessive-compulsive tendencies, was wheeled into the operating room. Electrodes were implanted at nine different sites in deep regions of his brain, and three months were allowed to pass after the surgery to allow for healing (Figure 1.2). Initially stimulation was delivered to all nine electrodes in turn. However, only the electrode implanted in the septum produced pleasurable sensations. When Patient B-19 was finally allowed free access to the stimulator, he quickly began mashing the button like an eight-year-old playing Donkey Kong. According to the paper,

> During these sessions, B-19 stimulated himself to a point that, both behaviorally and introspectively, he was experiencing an almost overwhelming euphoria and elation and had to be disconnected despite his vigorous protests.

So, not to put too fine a point on it, Heath's patient responded just as Olds and Milner's rats had. Given the chance, he would stimulate his pleasure circuit to the exclusion of all else.

Lest anyone think that it is only men—creatures of inherently base urges—who would respond in this manner, another recorded case, performed by a different group, involved a woman who received an electrode implant in her thalamus, an adjacent deep brain structure, to control chronic pain. This technique has proven effective for some patients whose severe pain is not well controlled by drugs. However, in this patient the stimulation spread to nearby brain structures, producing an intense pleasurable and sexual feeling:

> At its most frequent, the patient self-stimulated throughout the day, neglecting her personal hygiene and family commitments. A chronic ulceration developed at the tip of the finger used to adjust the amplitude dial

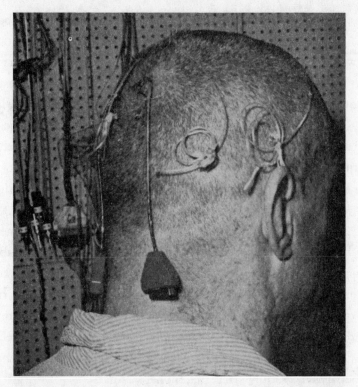

Figure 1.2 A patient of Dr. Robert Galbraith Heath with chronically implanted electrodes, one of which activated the medial forebrain bundle passing through the septum, a key part of the pleasure circuit. From Robert G. Heath, "Depth recording and stimulation studies in patients," in Arthur Winter, ed., *The Surgical Control of Behavior* (Springfield, Il.: Charles C. Thomas, 1971), 24. Reprinted with permission from Charles C. Thomas.

and she frequently tampered with the device in an effort to increase the stimulation amplitude. At times she implored her family to limit her access to the stimulator, each time demanding its return after a short hiatus.[3]

Back to Patient B-19: Before his brain stimulation began, he was shown a "15 min long 'stag' film featuring sexual intercourse and related activities between a male and female." Not surprisingly, he was sexually indifferent to this material and even a bit angry about being made to view it. Following pleasure circuit self-stimulation, however, he readily agreed to re-view the film ". . . and during its showing became sexually aroused, had an erection and masturbated to orgasm." All this in the decidedly unsexy environment of the lab. So, with the patient starting to exhibit heterosexual tendencies, what were the experimenters to do? Would he ever have an actual sexual relationship with a woman? After careful consideration of all the options and with the well-being of the patient foremost in their minds, Drs. Heath and Charles E. Moan made a sober medical and scientific decision: Upon getting approval from the attorney general of the state of Louisiana, they hired a hooker to come to the lab at Tulane and attempt to seduce him. She succeeded—they had sexual intercourse. The concluding sentence to the lengthy, overly descriptive paragraph describing their two-hour-long sexual encounter reads, "Then, despite the milieu and the encumbrance of the electrode wires [poor B-19 was attached to an EEG machine the whole time], he successfully ejaculated [in her vagina]."

Did Patient B-19 actually become heterosexual? Following discharge from the hospital, he had a sexual relationship with a married woman for several months, much to the delight of Drs. Moan and Heath, who found this development quite encouraging. His homosexual activity was reduced during this period, but did not stop completely: He still liked to turn tricks with men to earn money. Long-term follow-up information was not available. Writing in the discussion section of their scientific report, Moan and Heath were enthusiastic about the prospects for this therapy: "Of central interest in the case of B-19 was the effectiveness of pleasurable stimulation of new and more adaptive sexual behavior." While it's clear that Patient B-19 found the brain stimulation to be intensely pleasurable, I'm not convinced that he truly became heterosexual, even temporarily. It should also be cautioned that this report concerns only a single individual, not a population (with a control group).

This study is morally repugnant on many different levels—the profound arrogance of attempting to "correct" someone's sexual orientation, the medical risk of unjustified brain surgery, the gross violations of privacy and human dignity. Fortunately, homosexual conversion therapy with brain surgery and pleasure center stimulation was soon abandoned. Stepping back a bit, what we are left with, from this and a handful of other studies, is an appreciation of the immense power of direct electrical stimulation of the brain's pleasure circuitry to influence human behavior, at least in the near term.

~~~~~

Let's now take a minute to consider some important details of the pleasure circuit. I hesitate to burden you with neuroanatomy, but just a smidgen will go a long way in explaining how we experience pleasure. We'll use the rat as an example, which is appropriate because the anatomy of the rat's pleasure circuit is very similar to that of our own (Figure 1.3). When neurons in the region called the ventral tegmental area (VTA) are active, brief electrical impulses (called spikes) race from their cell bodies (located in the VTA proper) along long, thin information-sending fibers called axons. The axons have specialized structures at their endpoints called axon terminals. Some of the axon terminals of the VTA neurons are located some distance away in a region called the nucleus accumbens. When the traveling electrical spikes reach the axon terminals, they trigger the release of the neurotransmitter dopamine, which is stored in the terminals in tiny membrane-bound blobs called vesicles. When a spike enters the axon terminal, it initiates a complex series of electrical and chemical events that result in the fusion of the vesicle membrane with the membrane of the axon terminal, thereby causing the contents of the vesicle, including the dopamine neurons, to be released into a narrow fluid-filled space surrounding the axon terminal called the synaptic cleft. The dopamine molecules then diffuse and bind to specialized dopamine receptors on their target neurons, initiating a series of chemical signals therein (Figure 1.4).

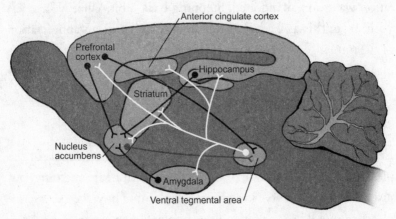

Figure 1.3 The pleasure circuit in the brain of a rat. This view shows a section through the middle of the rat brain, oriented so that the nose is at the left and the tail at the right. The central axis of the pleasure circuit is the dopamine-containing neurons of the ventral tegmental area (VTA) and their axons, drawn in white, which project to the nucleus accumbens. These VTA neurons also send their dopamine-releasing axons to the prefrontal cortex, the dorsal striatum, the amygdala, and the hippocampus. The VTA neurons receive excitatory drive from the prefrontal cortex and inhibitory drive from the nucleus accumbens. Illustration by Joan M. K. Tycko.

Neurons of the VTA also send dopamine-releasing axons to other brain regions, including the amygdala and the anterior cingulate cortex, which are emotion centers; the dorsal striatum, involved in some forms of habit learning; the hippocampus, involved in memory for facts and events; and the prefrontal cortex, a region

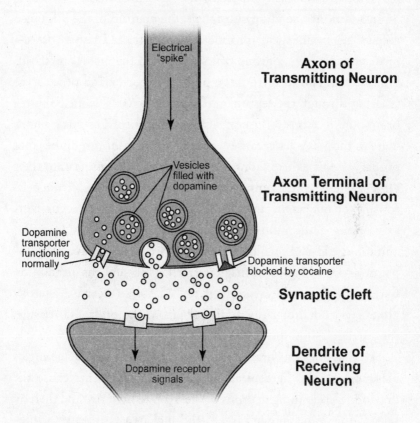

Figure 1.4 A synapse that uses the neurotransmitter dopamine. Dopamine is packed into membrane-bound vesicles in the presynaptic (information-transmitting) neuron. When electrical spikes traveling down the axon reach the axon terminal, they trigger the fusion of dopamine-containing vesicles with the presynaptic membrane, causing dopamine to spill out into the fluid-filled synaptic cleft. Dopamine released into the cleft can bind dopamine receptors on the dendrite of the postsynaptic (information-receiving) neuron, thereby exerting its effects, or undergo reuptake into the axon terminal through a dopamine transporter, where it will be recycled for later use. Cocaine and amphetamines block this reuptake process, causing dopamine to linger in the cleft and thereby activate dopamine receptors more effectively. Illustration by Joan M. K. Tycko.

that controls judgment and planning (and that is particularly expanded in humans as compared with other mammals).

In addition to sending out signals, the neurons of the VTA also receive electrochemical information from other brain regions—most notably from a group of axons, called the medial forebrain bundle, that run to it from the prefrontal cortex and other areas (passing through the septum and the thalamus). The medial forebrain bundle axons release the excitatory neurotransmitter glutamate in the VTA. This causes the VTA neurons to fire spikes that propagate down their axons and then release dopamine onto their targets. The VTA dopamine neurons also receive signals from neurons in the nucleus accumbens. However, the nucleus accumbens axons release the inhibitory neurotransmitter GABA (gamma-aminobutyric acid), which silences VTA neurons, preventing dopamine release.[4] The neurons in the nucleus accumbens, in addition to receiving dopamine axons from the VTA, also receive excitatory glutamate-containing fibers directly from the prefrontal cortex, the amygdala, and the hippocampus.

So what does this brain circuit diagram mean for the feeling of pleasure? Here's the central insight: Experiences that cause the dopamine-containing neurons of the VTA to be active and thereby release dopamine in their targets (the nucleus accumbens, the prefrontal cortex, the dorsal striatum, and the amygdala) will be felt as pleasurable, and the sensory cues and actions that preceded and overlapped with those pleasurable experiences will be remembered and associated with positive feelings. When Olds and Milner implanted electrodes to directly activate the pleasure circuit of rats, they placed them in positions that effectively stimulated the medial forebrain bundle, the axons that excite the dopamine neurons of the VTA. In fact, the electrode locations that produced the strongest pleasure (as determined by the frequency and duration of lever pressing the rats would perform) were those that most effectively activated the dopamine neurons of the VTA. Likewise, Moan and Heath's Patient B-19 and the handful of other humans who have received pleasure from direct brain stimulation all had electrode placements that produced excitation of the VTA dopamine neurons.[5]

How does activation of VTA dopamine neurons produce pleasure? Several lines of evidence indicate that catching a buzz from direct activation of the brain's pleasure circuitry with electrodes depends specifically upon release of dopamine in the target regions of the VTA neurons. Dopamine is released in these areas and diffuses across the synaptic cleft (the fluid-filled gap between neurons) before it binds and activates receptors on the target cells to exert its effects (Figure 1.4). After it has been released, dopamine doesn't just diffuse away; most of it is taken back up into axon terminals through the action of a protein called the dopamine transporter. The recycled dopamine is then repackaged into storage vesicles to be released again later. This means that drugs that block dopamine transporters will augment and extend the natural action of dopamine on the dopamine receptors of the target neurons, producing a more intense and longer-lasting signal.

When moderate doses of drugs that block the dopamine transporter, like amphetamines and cocaine, are given to rats, they increase the amount of lever pressing for medial forebrain bundle stimulation. This effect is most easily observed if the strength of the electrical stimulator is turned down a bit so it is not already producing maximum pleasure. Conversely, either drugs that block dopamine receptors (which are called dopamine receptor antagonists or neuroleptics) or surgical destruction of the dopamine cells of the VTA will cause a rat that has been engaging in vigorous lever pressing for pleasurable stimulation to stop.

Some psychoactive drugs work, at least in part, by hijacking the pleasure circuit—they artificially increase the effects of dopamine release from VTA neurons. (We'll get into the details of this in the next chapter.) Like humans, rats will self-administer certain drugs if given the chance. If you hook up a rat to a Skinner box in which lever presses deliver cocaine or amphetamines (either intravenously or directly into the brain), it will press the lever repeatedly. It will do so even if it has to work very hard to achieve the desired results—that is, even if a hundred lever presses are required to deliver a single tiny injection of drug. Just like Olds and Milner's rats with electrodes implanted in their pleasure circuits, and just like humans with implanted electrodes given unlimited access to brain pleasure circuit stimulation, rats allowed to self-administer cocaine and amphetamines will ignore food, water, sex, their personal hygiene, and even their young offspring in favor of the drug. With such behavior these "rat addicts" are a horrifyingly accurate reflection of the ruined lives of human drug addicts.

~~~~~

We've discussed what follows artificial activation of the dopamine-using pleasure circuit using electrodes or certain psychoactive drugs: intense pleasure that can, in some cases, become the basis for a profound and self-destructive addiction. But what happens under the opposite circumstances, when the dopamine neurons in the pleasure circuit are damaged or destroyed, thereby reducing the amount of dopamine released and thus dialing down the rheostat of pleasure?

Parkinson's disease is a neurological disorder that typically afflicts people over fifty and is more common in men than women. Its most striking symptoms involve the control of movement, and include tremor (in the hands, legs, jaw, and face), stiffness, slowness, and problems with balance and coordination. Parkinson's is chronic and progressive, usually starting with a subtle tremor but often progressing in a downward slide, resulting in difficulties with basic functions such as walking, talking, and the physical actions of eating (swallowing and chewing). Some cases run in families, but most do not. While the *ultimate* cause of Parkinson's disease is unknown, the *proximate* cause is well described: It results from a loss of dopamine-containing cells in the brain, particularly in two regions, the substantia nigra and the VTA—an integral part of the pleasure circuit. At present, there is no cure for Parkinson's disease, but a number of therapies are available to treat the symptoms by increasing dopamine action to compensate for the loss of dopamine-containing neurons. One is a drug called levodopa, a chemical precursor of dopamine. When levodopa is taken up into the surviving dopamine neurons, it causes them to make more dopamine and then to boost their release of dopamine. Obviously, this depends upon there being a sufficient number of surviving dopamine neurons—if they are all dead, then levodopa therapy will fail. Another therapy is a class of drug called dopamine receptor agonists. These drugs bind to dopamine receptors and activate them, thereby mimicking the actions of dopamine.[6] Levodopa and dopamine receptor agonists are often given together, and in many cases can significantly relieve the tremor and other movement problems, providing years of symptomatic relief. Unfortunately, however, as Parkinson's progresses, more and more dopamine neurons die, and as a consequence these drugs have to be given at higher and higher doses to control the symptoms.

In his original 1817 description of the "shaking palsy" that came to bear his name, James Parkinson classified it as a pure movement disorder that left "the senses and intellect . . . uninjured." However, that assessment has turned out not to be true, and likely resulted from the surprising fact that Parkinson examined only a single shaking palsy patient in his medical office—the other five cases that formed the basis of his essay were merely observed and sometimes interviewed as he walked around the streets of London.[7]

As early as 1913 it had become clear to some neurologists that Parkinson's disease also involved symptoms of cognition and mood and that these symptoms often preceded the onset of movement disorders. On average, Parkinson's patients tend to be introverted, rigid, stoic, slow to anger, and generally uninterested in seeking out novel experiences. They use alcohol, tobacco, and other psychoactive drugs at a much lower rate than the general population. Their personalities and actions are, in fact, the polar opposite of typical drug addicts, who are more likely to be impulsive, quick-tempered novelty seekers. Given that profile, it was quite unusual when case reports started appearing showing an unusually high incidence of pathological gambling among Parkinson's patients.

In January 2001 a sixty-four-year-old man was admitted to an outpatient treatment facility for problem gamblers in northern Italy. He had lost the equivalent of $5,000 playing slot machines over a three-year period, causing him to become estranged from his wife and to move in with his elderly mother. He had been diagnosed with Parkinson's disease twelve years earlier and had been placed on dopamine replacement therapy, consisting of a combination of levodopa and dopamine receptor agonists. The psychiatrists at the clinic performed cognitive-behavioral therapy (which is often effective for pathological gamblers) and prescribed antidepressants (serotonin-specific reuptake inhibitors, SSRIs). The treatment was unsuccessful, and the patient soon dropped out of the program and resumed his compulsive gambling. When he finally returned to the clinic in September 2002, the psychiatrists in charge had an insight—they asked the patient's daughter to surreptitiously monitor her father's intake of drugs. What they found was that, acting on his own, he had significantly increased his prescribed dose of both levodopa and dopamine receptor agonists. When confronted, he readily admitted increasing the dose and stated that he enjoyed the euphoric mood that accompanied it. (He liked the gambling too—he was just concerned about losing all his money.) When the drug doses were reduced to the prescribed level, his gambling stopped within a few days. In recent years, there have been a flood of similar case reports in the medical literature.[8]

Analysis of these reports revealed an interesting finding. Among patients treated solely with levodopa, the incidence of compulsive gambling was very low. (It's about 1 percent in the general population.) However, among Parkinson's patients treated with dopamine receptor agonists, there was a remarkable 8 percent incidence of pathological gambling. Gambling was in fact only the most common manifestation of a broad range of impulse control disorders. A small but significant population of these patients started compulsively eating, shopping (or shoplifting), and having risky sex—all behaviors that are highly atypical for the normal Parkinson's sufferer. In almost all cases, the impulse control disorder began shortly after an increase in the dose of dopamine receptor agonist and could be terminated by reducing the dose.

The best explanation for these findings is that in untreated Parkinson's disease, chronically low levels of dopamine result in a dialing down of the pleasure/reward circuit and a disinclination to seek novel experiences, and are associated with a reduced risk of addiction. In contrast, in some Parkinson's patients treated with high doses of dopamine receptor agonists, the level of dopamine action, in both the pleasure circuit and associated structures, is high, thereby dialing up the function of the pleasure circuit. This confers increased vulnerability to impulse control disorders and addiction.

~~~~~

To this point we've considered how the pleasure circuit can be artificially activated—hijacked by drugs or zapped by implanted electrodes. We've also examined the opposite case, in which the function of the pleasure circuit (and related structures) is attenuated in Parkinson's disease. While these stories are illuminating and help establish the existence of the pleasure circuit, we must now consider how the pleasure circuit functions naturally, in the healthy state, in the absence of artificial manipulations. Of course, the pleasure circuits of the brain have not evolved just to be activated by implanted electrodes or drugs. We must experience basic behaviors such as eating, drinking, and mating as pleasurable (rewarding) in order to survive and procreate. This consideration is not unique to humans. Indeed, rudimentary pleasure pathways appear quite early in evolutionary history. Even the soil-dwelling roundworm *C. elegans*, which is a millimeter long and has only 302 neurons in its entire body, has some basic pleasure circuitry. These worms typically feed on bacteria and are very good at following odor cues to find clumps of them. However, when a group of eight key neurons containing dopamine are silenced, the worms are mostly indifferent to this favorite food source (even though they can still detect odors). To anthropomorphize, the worms just don't seem to find eating bacteria to be that much fun anymore. This indicates that some aspects of the biochemistry of pleasure appear to have been conserved through hundreds of millions of years of evolution. In both modern roundworms like *C. elegans* (which follow an ancient body plan) and humans, dopamine-containing neurons occupy a central position in the pleasure circuit. This evolutionary conservation, from worms to humans, speaks to the central role of pleasure in the development of behavior.

In humans, rats, and other mammals, the reward circuit is much more complex as it is interwoven with brain centers involved in decision making, planning, emotion, and memory storage. When we find an experience pleasurable, it sets in motion several processes with different time courses: (a) We like the experience (the immediate sensation of pleasure); (b) we associate both external sensory cues (sights, sounds, odors, etc.) and internal cues (our own thoughts and feelings at the time) with the experience, and these associations allow us to predict how we should behave to repeat it; and (c) we assign a value to the pleasurable experience (from a little to a lot), so that in the future we can choose among several pleasurable experiences and determine how much effort we are willing to expend and risk we are willing to take in order to get them. (As my old friend Sharon expressed it, "I never met a man who excited me as much as a baked potato with sour cream.")

~~~~~

Human societies strictly regulate pleasurable activities, and most have a concept of vice that's applied to unregulated indulgence in food, sex, drugs, or gambling. Using a brain scanner, it has now become possible to observe activation of the brain's pleasure circuitry in humans. Not surprisingly, this circuit is activated by "vice" stimuli: orgasm, sweet and fatty foods, monetary reward, and some psychoactive drugs. What's surprising is that many behaviors that we consider virtuous have similar effects. Voluntary exercise, certain forms of meditation or prayer, receiving social approval, and even donating to charity can all activate the human pleasure circuit. There's a neural unity of virtue and vice—pleasure is our compass, no matter the path we take.

As we shall discover, the evidence for a general neurobiological model of pleasure is compelling and is only growing stronger as more research is done. How, then, should we think about the pleasures—both virtues and vices—that animate our lives? That wonderful meal, the blissful feeling of connectedness during prayer, that night of great sex, the "runner's high" from a Saturday-morning workout, a hilarious tipsy night at the bar with friends— are those all reducible to medial forebrain circuit activity and dopamine surges? Well, yes and no. Yes, in the sense that there seems to be a neural rheostat of reward involving the medial fore-brain dopamine circuit that's engaged by almost everything we find pleasurable.[9] No, in the sense that the activity of the pleasure circuit in isolation results in a lifeless pleasure lacking color and depth. What makes pleasure so compelling is that through the interconnection of the pleasure circuit with other brain regions, we adorn it with memory, with associations and emotions and social meaning, with sights, sounds, and smells. A circuit-level model of pleasure shows us what is necessary but not sufficient. The transcendence and the texture emerge from the web of associated sensations and emotions that the pleasure circuit engages. Our task now shall be to explore how that all plays out in the brain.

CHAPTER TWO

STONED AGAIN

*The psychoactive drugs of my people are transcendent and sub-
lime, while those of yours are base, crude, and sinful. My drugs
give me wisdom and foster creativity and spiritual insight; yours
are merely feeble crutches that reveal your lack of fortitude and
willpower. They transform you into a lazy and repulsive creature.
They make you behave like a beast.*

~~~~~~

All cultures use drugs that influence the brain. They range from
mild stimulants like caffeine to drugs with potent euphoric effects, like morphine. Some carry a high risk of addiction, some do
not. Some alter perception, others mood, and some affect both. A
few can kill when used to excess. The specific attitudes and laws
relating to psychoactive drug use vary widely among cultures,
however, and as such reinforce a narrative—summarized above—
in which those drugs used by insiders are considered acceptable,
while the use of drugs by outsiders is condemned and confirms

their status as somewhat less than human. This attitude is particularly evident in the course of the building of empires, with its attendant subjugation of other cultures. Mordecai Cooke, an English naturalist, writing in 1860, was, for his time, unusually insightful in this regard:

> Opium indulgence is, after all, very un-English, and if we smoke our pipes of tobacco ourselves, while in the midst of the clouds, we cannot forebear expressing our astonishment at the Chinese and other who indulge in opium. . . . [A]ll who have a predilection for other narcotics than those which Johnny Englishman delights in, come in for his share of contempt.[1]

These ideas about the psychoactive drugs of other cultures are not just a curiosity of Victorian England but are widespread and persist to this day. Nearly a century after Cooke's observation, the American author and drug enthusiast William S. Burroughs expressed a very similar notion when he wrote in his novel *Naked Lunch*, "Our national drug is alcohol. We tend to regard the use of any other drug with special horror."

Before we begin to examine the biology of psychoactive drugs—before we open the hood, so to speak—let's consider a few examples of drug use by different cultures in a variety of historical periods. This will serve to calibrate our thinking by offering a somewhat broader view in attempting to formulate cross-cultural, biological explanations of drug use.

~~~~~~~~

Rome, AD 170. It's a fine time to be a Roman nobleman. The empire is large and thriving, and the military seems invincible. Marcus Aurelius, later to be called "the last of the five good Roman emperors," is on the throne, attended by Galen, the famous Greek physician. And opium is freely available. Marcus Aurelius is best known today for his *Meditations*, a classic work of Late Stoic philosophy that holds that the denial of emotion is the key to transcending the troubles and pains of the material world.[2] Perhaps it was easier to be a Stoic while stoned: The emperor was a notorious opium user, starting each day, even while on military campaigns, by downing a nubbin of the stuff dissolved in his morning cup of wine. Writings by Galen suggest that Marcus Aurelius was indeed an addict, and his accounts of the emperor's brief periods without opium, as occurred during a campaign on the Danube, provide an accurate description of the symptoms of opiate withdrawal.

Opium, prepared from the poppy plant, *Papaver somniferum*, was in use long before the time of imperial Rome. Evidence from the archeological record places it in Mesopotamia (present-day Iraq) around 3000 BC. Opium was widely consumed—either by being eaten, dissolved in wine, or inserted in the rectum—for both medical and ritual purposes by the ancient Egyptians and by the Greeks soon thereafter. The Ebers Papyrus, an ancient Egyptian medical text from the year 1552 BC, even recommends opium as an aid to help small children sleep. One method to achieve this was to smear the drug on the nipples of the nursing mother.

While Galen popularized the use of opium, and it became widely consumed among the Roman nobility, it was not until several years later, during the reign of Septimus Severus, that the last legal restrictions on distribution of the substance were removed. This triggered the widespread adoption of opium as a Roman recreational drug. In the years that followed, the poppy plant became a symbol of Rome, stamped upon its coins, inscribed upon its temples, and woven into its religious practice. By the census of

AD 312, opium could be procured in 793 different shops in Rome, and the taxes on its sale constituted a substantial fraction of total receipts for the emperor.

~~~~~

County Derry, Ireland, 1880. During the 1830s Ireland was awash in alcohol, much of it produced locally in response to high alcohol import taxes imposed by the ruling British government. While many locals were assiduously distilling illegal *poitín* from potatoes or malted barley, a backlash against alcohol was also growing. The leading figure in this Irish temperance movement was a Catholic priest named Father Theobald Mathew, who in 1838 established the Total Abstinence Society. Its credo was simple: People who joined did not merely promise to consume in moderation, but took The Pledge, a commitment to complete abstinence from alcohol from that day forward. This simple approach was remarkably effective: In a single day more than twenty thousand drinkers were reported to have taken an oath of total abstinence at Nenagh, in County Tipperary. In fact, it is estimated that by 1844 roughly three million people, or about half the adult population of Ireland, had taken The Pledge.

Not surprisingly, some people looked for a way to keep to the letter of The Pledge while violating its spirit. One of these was a Dr. Kelly of Draperstown, County Derry, who realized that as a nonalcoholic tipple, ether filled the bill nicely. Ether is a highly volatile liquid that may be produced by mixing sulfuric acid with alcohol, as discovered by the German chemist Valerius Cordus around 1540.[3] The inhalation of ether vapors leads to effects that range from euphoria to stupor to unconsciousness. In fact, ether was the first drug ever to be used for general anesthesia when in 1842 Dr. Crawford Long of Jefferson, Georgia, employed it during the removal of a tumor from the neck of a patient. Dr. Long had been introduced to ether as a recreational drug during "ether frolic" parties while a medical student at the University of Pennsylvania and had the insight to imagine its practical use during surgery.

Dr. Kelly, desperate to become intoxicated while maintaining The Pledge, realized that not only could ether vapors be inhaled, but liquid ether could be swallowed. Around 1845 he began consuming tiny glasses of ether, and then started dispensing these to his patients and friends as a nonalcoholic libation. It wasn't long before it became a popular beverage, with one priest going so far as to declare that ether was "a liquor on which a man could get drunk with a clean conscience." In some respects ingesting ether is less damaging to the system than severe alcohol intoxication. Ether's volatility—it's a liquid at room temperature but a gas at body temperature—dramatically speeds its effects. Dr. Ernest Hart wrote that "the immediate effects of drinking ether are similar to those produced by alcohol, but everything takes place more rapidly; the stages of excitement, mental confusion, loss of muscular control, and loss of consciousness follow each other so quickly that they cannot be clearly separated." Recovery is similarly rapid. Not only were ether drunks who were picked up by the police on the street often completely sober by the time they reached the station, but they suffered no hangovers.

Ether drinking spread rapidly throughout Ireland, particularly in the north, and the substance soon could be purchased from grocers, druggists, publicans, and even traveling salesmen. Because ether was produced in bulk for certain industrial uses, it could also be obtained quite inexpensively. Its low price and rapid action meant that even the poorest could afford to get drunk several times a day on it. By the 1880s ether, distilled in England or Scotland, was being imported and widely distributed to even the smallest villages. Many Irish market towns would "reek of the mawkish fumes of the drug" on fair days when "its odor seems to cling to the very hedges and houses for some time." In 1891, Norman Kerr, writing in the *Journal of the American Medical Association*, painted a vivid picture of pervasive ether intoxication:

> Sturdy Irish lads and beautiful Irish lasses, brimful of Hibernian wit, . . . are slaves to ether drunkenness. The mother may be seen with her daughters and maybe a neighboring Irishwoman or two at a friendly ether "bee." The habit has become so general that small shopkeepers treat the children who have been sent to purchase some article, with a small dose of ether, and schoolmasters have detected ether on the breaths of children from 10 to 14 (or even younger) on their arrival at school.[4]

It is interesting to note that, even at the peak of the Irish ether-drinking craze, the possession, sale, and private use of ether remained legal. The first attempt to control the problem involved adulterating industrial ether with naphtha, which has an odor and taste even more offensive than ether itself. This was an utter failure—people just blended it with sugar and spices to mask the taste, held their noses, and tossed it back. Ether drinking in Ireland was finally curtailed in 1891 when the British government classified ether as a poison and enforced strict controls on its sale and possession, thus dramatically restricting its distribution and use. The practice lingered for a few years longer but appeared to be completely abolished by the 1920s.

Cheap, quick, and no hangover afterward? No wonder ether was so popular. However, before you head out the door to score some, it's worth mentioning a few of the downsides. These include a truly awful smell and taste, coupled with a strong burning sensation while the foul stuff is going down. Plus, it makes you drool like a Saint Bernard dog on a hot summer day, not to mention stimulating truly monumental burps and farts. These aren't normal emissions— they are laden with highly flammable ether vapors. You can imagine what happened when an ether drinker would light up a pipe and belch or sit down by an open fire and break wind. Severe burns at either end of the alimentary canal were a common hazard.[5]

~~~~~

Iquitos, Peru, 1932. Emilio Andrade Gomez was born in the Peruvian Amazon, the son of a white father and an Amerindian mother. In 1932, at the age of fourteen, he was given the herbal hallucinogenic drink called ayahuasca by local shamans in order to recover his strength following a period of illness.[6] He saw visions that the shamans explained were revelations that he was chosen by the plants in the ayahuasca brew to receive knowledge from them. He was to learn traditional medicine and become a shaman himself. This was an elaborate and extended process that required him to live in near isolation in the jungle for a period of three years. During this time he was provided a strict traditional diet, consisting mostly of plantains and fish. He could eat some jungle fowl, but only the left breast—no other portion of the meat was allowed. Alcohol and sexual contact were strictly prohibited. His food was prepared and delivered to him by either a young girl or a postmenopausal woman, and whatever portion remained uneaten was carefully collected and destroyed so that no other man or animal might consume it.

During the period of this ritual diet, Don Emilio wandered in the jungle to study plants, and about every two weeks he would prepare and drink ayahuasca. The substance revealed to him a supernatural world, filled with spirits of both malign and benevolent character. The malign spirits were those of evil shamans who sought to pierce his body with magical darts that would sap his strength or even kill him. The benevolent spirits were sometimes those of good shamans, but more commonly were the spirits of jungle plants. In his ayahuasca dreams of those years, Don Emilio learned about sixty different songs from the plant spirits. When the initiation period was complete, he began his own practice of medicine, under the tutelage of an elder shaman. He used the *icaros*, the magic songs taught by the plants, to reinforce the effectiveness of his own herbal preparations and thereby help others to cure particular diseases, attract game or fish, repel the attack of evil shamans, or win the love of another.

It is not known when or exactly where the ritual use of ayahuasca began, but it is likely to have been hundreds of years before

Figure 2.1 Don Emilio Andrade Gomez preparing ayahuasca near Iquitos, Peru, in 1981. This photo shows the finished product being decanted. Photograph by Dr. Luis Eduardo Luna. Used with permission.

European colonization. Ayahuasca first became known to Europeans following the 1851 Amazon expedition of the English botanist Richard Spruce, who observed it being consumed by the Tukano people living on the Rio Uaupés in Brazil. Since then, ayahuasca use has been found to be widespread among Amerindian groups in the upper Amazon basin, with practitioners in present-day Peru, Ecuador, Brazil, and Colombia.

The recipe for ayahuasca varies across different groups, but the most common preparation is as follows: A shaman gathers a particular type of liana vine (*Banisteriopis caapi*) and prepares about thirty pieces of the crushed stem, each about a foot long, in about four gallons of water. To this are added about two hundred leaves of the chacruna bush (*Psychotria viridis*), following which the woody, leafy stew is boiled slowly for about twelve hours, until the volume of the liquid is reduced to about a quart and has become an oily, gritty brown syrup (see Figure 2.1). This is sufficient for about twelve foul-smelling doses. About a half hour after swallowing the brew, hallucinations begin, and typically last for three to six hours. These are primarily visual, sometimes auditory, and frequently terrifying. For most, it's a rather inward-turning, paranoiac, fearful experience, but one that can yield insight and self-knowledge. Ayahuasca drinking is almost always accompanied by vomiting, and examination of the vomit is said to yield clues as to the efficacy and nature of the treatment.

While plant-based psychoactive drugs are common throughout the world, ayahuasca is unusual in that its action specifically requires the activity of two different classes of compound from two different species of plant. The hallucinations are the result of the dimethyltryptamine (DMT) in the chacruna leaves. DMT is a molecule with a chemical structure similar to that of the well-known synthetic hallucinogen LSD. However, unlike LSD, DMT taken by mouth is completely broken down in the digestive tract by the enzyme monoamine oxidase, and so none reaches the brain to produce its psychoactive effects. The liana contains a group of related beta-carboline compounds, one of which is called harmaline. Harmaline and its relatives are potent monoamine oxidase inhibitors. When taken alone at the doses typically found in ayahuasca preparations, harmaline does not lead to hallucinations. (It does, however, produce a strong tremor and discoordination of movement.) But when liana and chacruna extracts are swallowed together, as in ayahuasca, the harmaline blocks the action of monoamine oxidase in the gut, allowing the DMT from the chacruna to escape breakdown and reach the brain. This is an interesting finding, because it suggests that the Amerindian discoverers of the ayahuasca preparation were unlikely to have simply stumbled upon the properties of the mixture during food preparation. It's more likely that ancient traditional healers of the Amazon basin had made a systematic study of the effects of specific combinations of plant extracts.

~~~~~

Berkeley, 1981. When I was in college, I knew some guys who had devised a unique and very dangerous form of Friday-night recreation. After each had consumed about ten beers, they would gather around a huge fishbowl that had been filled about half full with many types of prescription pills, mostly psychoactive drugs. Quaaludes, Valium, amyl nitrite, Dexedrine, Percodan, and Nembutal were all in the mix, as were antihistamines, laxatives, over-the-counter painkillers, and God knows what else. The game was to reach into the bowl and randomly grab two different pills, make note of their color and shape, and swallow them immediately. Then, while waiting for the buzz to kick in, each celebrant would open the huge book next to the fishbowl (the *Physicians' Desk Reference*, which lists all the pills produced by drug companies, together with a set of color photographs to aid in their identification) to learn about what he had just ingested, reading aloud to the group. I have a strong memory of a huge, shaggy blond kid (imagine Jeff Spicoli from *Fast Times at Ridgemont High* writ large) working his way doggedly through a section on potential side effects and mumbling in his surfer-stoner monotone, "Whoa . . . cerebral hemorrhaging . . . cool."

~~~~~~

So what can we conclude from these four examples? First, psychoactive drugs can be used in many different social contexts: as medicine, as religious sacrament, as pure recreation, or to define oneself as part of a subgroup (elite, outsider, rebel, etc.). Second, these contexts can change and overlap. Opium in ancient Rome and Quaalude pills in the United States in the 1980s were both initially used for medicinal purposes but rapidly became mostly recreational; ayahuasca in the Amazon basin is used as both a medicine and a religious sacrament. Third, religious edicts and laws of the state can have a profound effect on drug use, often in unexpected ways. The nineteenth-century Irish ether-drinking craze resulted in large part from high taxes on ethanol imposed by the British government, combined with the influence of Father Mathew's Total Abstinence movement. The use of opium didn't really explode in ancient Rome until Emperor Septimus Severus lifted the last of the restrictions on its sale, in large part to increase tax revenues and thereby fund his military exploits. However, the most important lesson to take from these examples is the simplest one: Across cultures and over thousands of years of human history, people have consistently found ways to alter the function of their brains, while cultural enforcers such as governments and religious institutions have sometimes, but not always, sought to regulate the use of these substances.

Lord Byron, the British romantic and satiric poet of the early nineteenth century, wrote, "Man, being reasonable, must get drunk; the best of life is but intoxication." While Byron was describing the effects of alcohol, the larger truth applies to psychoactive drugs generally. Because most are derived from plant extracts (cannabis, cocaine, caffeine, ibogaine, khat, heroin, nicotine) or from simple recipes applied to plants (alcohol, amphetamines) or fungi (mescaline), they are widely available and widely used.

In fact, intoxication with psychoactive drugs is not an exclusively human proclivity. Animals in the wild will also voluntarily and repeatedly consume psychoactive plants and fungi. Birds, elephants, and monkeys have all been reported to enthusiastically seek out fruits and berries that have fallen to the ground and undergone natural fermentation to produce alcohol. In Gabon, which lies in the western equatorial region of Africa, boars, elephants, porcupines, and gorillas have all been reported to consume the intoxicating, hallucinogenic iboga plant (*Tabernanthe iboga*). There is even some evidence that young elephants learn to eat iboga from observing the actions of their elders in the social group. In the highlands of Ethiopia, goats cut the middleman out of the Starbucks business model by munching wild coffee berries and catching a caffeine buzz.[7]

But do we really know whether these animals *like* the psychoactive effects of the drug or are just willing to put up with them as a side effect of consuming a valuable food source? After all, fermented fruit is a tasty and nutritious meal. While it's hard to dissociate these motivations in animals, many cases suggest that the psychoactive effect is the primary motivator for consumption. Often, only a tiny amount of plant or fungus is consumed, so while its nutritional effect is minuscule, its psychoactive effect is large.

Perhaps the most dramatic example of nonnutritive animal intoxication is found among domesticated reindeer. The Chuckchee people of Siberia, who are reindeer herders, consume the bright red hallucinogenic mushroom *Amanita muscaria* as a ritual sacrament. Their reindeer also indulge. On discovering the mushrooms growing wild under the birch trees, they gobble them up and then stagger around in a disoriented state, twitching their heads repeatedly as they wander off from the rest of the herd for hours at a time. The active ingredient of the *Amanita* mushroom is ibotenic acid, a portion of which is converted in the body into another compound called muscimol—the substance that actually produces the hallucinations.[8] What's interesting about ibotenic acid is that only a fraction is metabolized in the body to form muscimol, while the rest—about 80 percent of that consumed—is passed in the urine. The reindeer have learned that licking ibotenic acid–laden urine will produce as much of a high as eating the mushroom itself. In fact, this drugged urine will attract reindeer from far and wide, and they will even fight over access to a particularly attractive patch of yellow snow. All of this has not gone unnoticed by the Chuckchee, who collect the urine of their *Amanita*-eating shamans for two reasons. The first is simple thriftiness: *Amanita* mushrooms are often scarce, so urine recycling can provide about five doses for the cost of one fresh one, albeit with a rather severe aesthetic penalty. The second is that the reindeer are just as enthusiastic about human *Amanita*-tainted urine as they are about their own, and so they can be effectively rounded up with a bit of the stuff sprinkled in a corral. Clearly, Siberian reindeer are not fighting over drugged urine for its nutritive value.[9]

～～～～

All this begs the question: Why is the use of psychoactive drugs so widespread? For simple pleasure? For brief spurts of energy? To reduce anxiety and foster relaxation and forgetting of one's troubles? To excuse behavior that would not otherwise be socially tolerated?

To stimulate creativity and explore new forms of perception? To augment ritual practice? The answer, of course, is all of the above. The psychiatrist Ronald K. Siegel holds that all creatures, from insects munching psychoactive plants to human children playing spinning games to get dizzy, have an inborn need for intoxication. He writes, "This behavior has so much force and persistence that it functions like a drive, just like our drives of hunger, thirst, and sex."[10] Do we, in fact, have an innate drive to alter the function of our brains? And if so, why?

~~~~~

Humans seek out drugs with a wide variety of psychoactive effects. A rough taxonomy would include stimulants, sedatives, hallucinogens, opiates, and drugs with mixed actions. The stimulants, which comprise a wide range of compounds that increase wakefulness and generally up-regulate mental function, include cocaine, khat, amphetamines (including Adderall and Ritalin), and caffeine. Stimulants generally have positive effects on mood, but can sometimes cause anxiety and agitation. The sedatives, of course, produce the opposite effects: They are calming and sleep-inducing, and cause discoordination and slow reaction times. Sedatives include alcohol, ether, barbiturates, the benzodiazepine tranquilizers (such as Halcion, Xanax, Rohypnol, and Ativan) and gamma-hydroxybutyrate (GHB). The hallucinogens (substances like LSD, mescaline, PCP, ketamine, and ayahuasca) have as their primary action the disruption of perception—distorting vision, hearing, and the other senses. They also produce complex alterations of cognition and mood, often involving an interesting sensation of "oneness with the universe." Opiates (including plant-derived compounds like opium, morphine, and heroin as well as synthetic opiates like OxyContin and fentanyl) are sedatives, but ones that deserve their own category because they produce a unique and potent euphoria (and capacity for pain relief), effects that are not shared by other sedatives with a different chemical action.

Of course, we know from our own experience that this kind of drug taxonomy is a blunt instrument. For example, cocaine is typically not considered a hallucinogen, but it can occasionally produce hallucinations at very high doses. Similar blurring of these categories of action comes when considering some of the world's most popular drugs. While alcohol at high doses is always a sedative (to the point where it can be lethal), at lower doses it has a stimulating effect, particularly in certain social contexts. This stimulation can lead to a range of outcomes—from animated conversation to sloppy karaoke to a bar fight. Nicotine has a complex and subtle psychoactive effect, with mixed actions of a stimulant, a sedative, and a mild euphoric. The popular club drug ecstasy (methylenedioxymethamphetamine, or MDMA) is both a stimulant and a weak hallucinogen that has the additional quality of inducing a sense of intimacy with others. Cannabis is a sedative but also has mild euphoric properties (more than nicotine but much less than heroin). Antidepressant drugs, like the serotonin-specific reuptake inhibitors (Prozac, Zoloft, Celexa, and others) or the dual-action antidepressants (such as Effexor), will lighten the mood of many people, whether or not they suffer from depression, but they don't easily fall into one of our five categories.

Perhaps the most important aspect of psychoactive drug action that is not captured by our simple taxonomy is social context. While a certain drug will always have the same chemical action, that action is influenced by one's ongoing brain state in ways that can modulate its effects. People who are given morphine for pain relief typically report a lot of pain abatement and only a mild euphoria. Others taking the same dose of morphine recreationally will report a much higher degree of euphoria. In a recent laboratory study, one group of experimental subjects was told that they were smoking an unusually potent strain of cannabis while another was told that they were smoking an unusually weak strain; in fact, both were given the same average-potency strain. The two groups smoked the same amount over roughly the same time period. The individuals who were told that they were smoking the super-potent cannabis not only reported significantly higher subjective ratings of euphoria (which is perhaps not so surprising), but they also had slower reaction times and greater discoordination of movement as measured with a precision reaching task. As the British addiction expert Griffith Edwards says, "A lot of what drugs do to the mind is in the mind."[11]

Many years ago this interaction between mental state, social context, and drug effect was brought home to me in dramatic fashion. Two college friends of mine who decided to take LSD together during finals week asked me to "babysit" them. One fellow, "Ned," had just finished his exams and was feeling as if the weight of the world had been lifted from his shoulders. The other (we'll call him "Fred") had one exam remaining a few days hence—in physics, a class in which he had struggled. Ned and Fred swallowed their doses, put on a Pink Floyd record (it was 1979, after all), and settled in on the sofa for their trips. Ned had a typical happy LSD experience, watching the colors shifting on the ceiling, laughing, and feeling generally blissed out. Fred, on the other hand, had a taste of hell. He became first withdrawn and then deeply paranoid. Soon he was weeping, thrashing around, and screaming about physics equations, Kirchhoff's circuit laws, and how he would never understand the weak nuclear force. He imagined being attacked by a monstrous Niels Bohr with blood-dripping fangs. It was the classic bad trip, and he never used LSD again.

~~~~~

Let's now return to the issue of pleasure. We know that experiences that cause the dopamine-containing neurons of the VTA to be active and thereby release dopamine in their target regions will be felt as pleasurable and that this process can be co-opted by direct activation using implanted electrodes. One simple hypothesis regarding drug use is that the various psychoactive substances that we humans seek out, whether they are stimulants, sedatives, opiates, hallucinogens, or drugs of mixed action, all activate the medial forebrain pleasure circuit. We've already discussed how the stimulants cocaine and amphetamines produce dopamine release from VTA neurons by blocking the reuptake of dopamine into axon terminals, thereby prolonging dopamine action in the VTA target regions and stimulating the pleasure circuit (Figure 1.4). What about other drugs that don't target the dopamine system? For example, we know that morphine and morphine-related drugs (like heroin and fentanyl) can produce potent euphoria yet have no direct effects on dopamine signaling.

In order to explain the action of opiate drugs, we'll need to pause for a bit and consider a few more general issues of drug action. Some psychoactive drugs have widespread effects. Both alcohol and ether, for example, act in nonspecific ways that impact many different chemical and electrical functions of neurons. Similarly, caffeine has a wide range of effects on neurons, and its action as a stimulant cannot be traced to a single one of these. However, most drugs, both natural and synthetic, work by acting on specific neurotransmitter systems of the brain. For example, cocaine and amphetamines block the reuptake of dopamine, benzodiazepine tranquilizers like Ativan work by binding and augmenting the natural action of receptors for the inhibitory neurotransmitter

GABA, and SSRI antidepressants like Prozac work by inhibiting the reuptake of released serotonin.

In many cases, the action of a drug is well known long before the identification of its target neurotransmitter system in the brain. When Sol Snyder and Candace Pert first demonstrated the biochemical function of the morphine receptor in 1973, this receptor had no known natural activator within the body. This was puzzling—it seemed unlikely that evolution would produce receptors in the brain that would be inactive until the animal consumed a particular species of poppy plant. Sure enough, two years later John Hughes and Hans Kosterlitz were the first to identify chemicals present in the brain that bind and activate morphine receptors. These natural analogs of morphine are called endorphins. Since then, a large family of opioid receptors with different biochemical actions has been discovered, accompanied by the description of a large number of endorphins. The role of the endorphin/opioid system is multifaceted, being implicated in a variety of functions including pain perception, mood, memory, appetite, and neural control of the digestive system.

A similar story has emerged for cannabis and tobacco. The main psychoactive ingredient in cannabis is the compound tetrahydrocannabinol, or THC, which binds and activates specific and unique receptors in the brain. These receptors, which are called CB1 and CB2, are naturally activated by the brain's own THC-like molecules. "Endocannabinoids" are the brain's own cannabis in the same sense that the endorphins are the brain's own morphine. To date, two endocannabinoids have been identified: 2-arachidonylglycerol and anandamide (the latter from the Sanskrit word *ananda*, meaning "bliss"). In tobacco the key psychoactive ingredient is nicotine, which activates a subset of the receptors for the endogenous neurotransmitter called acetylcholine.[12]

The euphoric portion of cannabis intoxication operates via

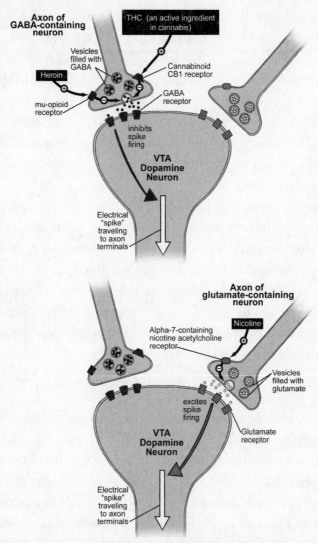

Figure 2.2 Heroin and related drugs (morphine, OxyContin, methadone) produce indirect activation of the pleasure circuit by reducing release of the inhibitory neurotransmitter GABA, resulting in disinhibition of VTA dopamine neurons. THC, the main psychoactive ingredient in cannabis, acts in a similar fashion (top panel). Conversely, nicotine evokes indirect activation of the pleasure circuit by increasing the release of the excitatory neurotransmitter glutamate, resulting in excitation of VTA dopamine neurons (bottom panel). Illustration by Joan M. K. Tycko.

an indirect signaling scheme (Figure 2.2, top). THC binds and activates CB1 endocannabinoid receptors on those presynaptic terminals that release GABA onto VTA dopamine neurons. When ongoing GABA release is reduced, VTA neurons are disinhibited and dopamine release is increased in the VTA target regions. Alcohol has an even more convoluted mode of action. It increases the secretion of both endorphins and endocannabinoids (through mechanisms that are not well understood) and thereby disinhibits VTA dopamine neurons.[13]

Nicotine produces a similar end result to morphine and THC, but uses the opposite logic: It binds and activates receptors on the glutamate-containing axon terminals that contact VTA dopamine neurons (Figure 2.2, bottom). When nicotine activates these specialized receptors (called alpha-7-containing nicotinic acetylcholine receptors), this action increases glutamate release, producing greater excitation of VTA neurons and, of course, increased dopamine release.

~~~~~

So we've seen that a whole range of psychoactive drugs up-regulate dopamine action in VTA target regions. Interestingly, these compounds span a broad segment of our drug taxonomy: from stimulants like cocaine and amphetamines to sedatives like alcohol to opiates like heroin and to drugs of mixed action like nicotine and cannabis. That's all very well, but couldn't dopamine release from VTA neurons merely be roughly correlated with the actions of these drugs but not have a direct role in their psychoactive effects? Probably not. When human subjects are placed in a brain scanner and given an intravenous hit of cocaine or amphetamines or heroin, strong activation of the VTA and dopamine release in the VTA target regions is seen, and these events peak precisely when the subjects report that their pleasure rush is strongest.

Our original hypothesis was that the various drugs that we humans seek out all activate the medial forebrain pleasure circuit. It seems to hold for the substances we've just discussed, but is the impulse universal? Do we always seek out drugs for pleasure? Well, no. For example, most hallucinogens—drugs like LSD, ayahuasca, and mescaline—don't activate the medial forebrain pleasure circuit. Many sedatives, like the barbiturates and benzodiazepines, fail to do so as well. The widespread cross-cultural (and even cross-species) drive to tamper with our brain function cannot be entirely accounted for by activation of the pleasure circuit. Pleasure is central to some—but not all—psychoactive drugs.

If some psychoactive drugs activate the pleasure circuit while others don't, what does this mean for the many users of these various drugs? Here's the news: Those psychoactive drugs that strongly activate the dopamine-using medial forebrain pleasure circuit (like heroin, cocaine, and amphetamines) are the very ones that carry a substantial risk of addiction, while the drugs that weakly activate the pleasure circuit (like alcohol and cannabis) carry a smaller risk of addiction.[14] Drugs that don't activate the pleasure circuit at all (like LSD, mescaline, benzodiazepines, and SSRI antidepressants) carry little or no risk of addiction. This pleasure gradient also correlates strongly with the willingness of animals to work for these drugs. Rats will perform hundreds of lever presses for a single tiny injection of cocaine but only a few for intravenous alcohol, and they are completely uninterested in performing lever presses for LSD or benzodiazepines or SSRIs.

Speaking of the "risk of addiction" associated with various psychoactive drugs is a common practice, and while the expression rolls off the tongue easily, it's admittedly a very crude construction. Sociocultural factors have a huge impact on the risk of addiction associated with a given drug. Obviously, if a drug isn't easily available to you, you're unlikely to use it. Thus legal drugs like alcohol and nicotine are broadly available, semilegal drugs like benzodiazepines, prescription amphetamines, and cannabis are somewhat less so, and illegal drugs like heroin and cocaine are difficult to procure and carry the most legal risk. In the United States many psychoactive drugs have become so inexpensive—a dose of LSD, ecstasy, or cannabis often costs the same as a large cappuccino at the mall—that purely economic considerations are minimized. Of course, the attitudes of one's peers, family, and faith will also have an impact on an individual's drug use.

For a particularly addictive drug like heroin, cocaine, or nicotine, it seems that the exact mode of intake is likewise crucial in determining its risk for addiction. For example, cocaine may be injected, smoked, snorted, or ingested, and it's well established that it is more addictive when it is smoked or injected than when it is snorted. This is the basis of the crack cocaine epidemic that devastated many communities in the late 1980s and that continues to be a scourge to this day. Smoked or injected cocaine is more addictive because it reaches the target neurons in the brain with a rapid onset, while snorted cocaine produces a pleasure rush that comes on somewhat more slowly. Ingested cocaine from coca leaf chewing—a traditional practice in the Andes Mountains of Peru and Bolivia—has an even slower time course of onset and is by far the least addictive route of administration.[15]

The situation regarding opiates is similar. When opium was prepared as a crude extract of the poppy plant, it could either be eaten, inserted rectally, or smoked. Of course, smoking produced a highly effective delivery of morphine to the brain and was much more addictive. However, a set of innovations in the nineteenth century set the stage for even more rapid delivery. First was the purification of morphine from opium by the German chemist Friedrich Sertürner in 1805. Second was the invention of the hypodermic syringe, which allowed for the injection of pure morphine solution

Figure 2.3 Cannabis delivered rectally would produce a psychoactive effect with very slow onset. Smoking coffee, on the other hand, might well deliver caffeine to the brain much more rapidly than drinking it. "Five minute comic" by Joey Alison Sayers (www.jsayers.com), from her book *I'm Gonna Rip Yer Face Off!* Used with permission.

into the bloodstream. The first widespread use of injectable morphine came during the American Civil War and was a boon for battlefield pain control in wounded soldiers. But there was a heavy price for such treatment, as many veterans, particularly on the Union side, returned home addicted to injected morphine. The third factor was an invention by the Bayer drug company, which in 1898 introduced heroin, a simple chemical derivative of morphine (with two added acetyl groups). Heroin had the advantage of being able to cross cell membranes more easily than morphine, producing an even faster onset of action in the brain and an even more pronounced pleasure rush.

It's important to note that even injecting heroin does not inevitably result in addiction. A recent study of drug use in the United

States estimates that about 35 percent of all people who have tried injected heroin have become heroin addicts. While that's a very high percentage relative to addiction rates of 22 percent for smoked or injected cocaine, about 8 percent for cannabis, and about 4 percent for alcohol, consider this shocking statistic: 80 percent of all the people who try cigarettes become addicted. In part, that remarkably high number reflects the fact that tobacco is legal and that the health and lifestyle penalties for smoking cigarettes, while significant, are much less than those related to heroin and often take many years to manifest.[16]

Why is cigarette smoking so addictive when its psychoactive effect is comparatively so subtle? The reason is that the cigarette is the Galil assault rifle of the nicotine delivery world: fast and reliable. Consider that while a heroin user injects a hit and feels a potent euphoric rush about fifteen seconds later, he is not going to inject again for many hours. The cigarette smoker, on the other hand, will typically take ten puffs from a single cigarette and will often smoke many cigarettes in the course of a day. Each puff will deliver nicotine to the pleasure circuit about fifteen seconds later, approximately the same delay as for intravenous heroin. So while a typical heroin addict may get two strong, rapidly delivered hits per day, the pack-a-day cigarette smoker will get two hundred weak, rapidly delivered hits per day. But why does the nearly instant delivery of a drug to the brain, as with smoking cigarettes or injecting heroin, carry a higher risk of addiction than slow delivery of the same drug, say, by chewing tobacco or eating opium?

One way to think about this is to consider that addiction is a form of learning. When someone uses a drug, associations are made between a particular act (injecting the drug or chewing the tobacco) and the pleasure that follows. Imagine that you have a dog that you're trying to train to come when called, using a tasty morsel of food as a reward. If you want to create a learned association, you'll call the dog, and when it comes you'll immediately give it the treat. Now imagine that instead of presenting the reward immediately (as with injected heroin), you wait thirty minutes and then offer the reward (as with ingested opium). In the latter case, the connection between the behavior (coming when called) and the reward is quite weak, and the association is less likely to be learned. The same dog-training analogy holds true for injected heroin (one big pleasure rush) and cigarette smoking (many tiny pleasure rushes). If you call the dog once a day, and then immediately reward its compliance with a ten-ounce steak, it will eventually learn to come when called. If you call the dog twenty times per day and immediately reward each correct behavior with a small chunk of meat, the dog will learn much more quickly. So when we smoke cigarettes, we are being very effective trainers of our inner dog, creating a strong association between puffing and pleasure.[17]

~~~~~

Addiction can be defined as persistent, compulsive drug use in the face of increasingly negative life consequences. Addicts typically risk their health, families, careers, and friendships as they pursue their drugs of choice. Addiction doesn't develop all at once, however, but proceeds in stages. While some drugs, like heroin, carry a high risk for addiction, not even heroin produces addiction with a single dose. When a drug user initially gets high on cocaine or heroin or amphetamines, the experience produces an intense euphoric pleasure and sense of well-being. It is repeated doses, particularly if strung closely together in a binge, that will often trigger the dark side. This is first manifested as drug tolerance: Soon after a binge, the drug user will need a higher dose to achieve the same level of euphoria, and if drug taking continues regularly, this tolerance will become greater and greater. As tolerance to the drug develops, so does dependence, which means that the addicted person not only needs more of the drug to get high, but also will feel bad in its absence. Dependence can be experienced as both mental symptoms (such as depression, irritability, or inability to concentrate in the absence of the drug) and physical ones (such as nausea, cramps, chills, and sweats).

In the later stages of addiction, users feel strong cravings for the drug, which are often triggered by drug-associated stimuli. A crack cocaine addict may be feeling relatively stable but will have an intense desire for the drug at the sight of a pipe. The cravings of an amphetamine addict who often gets high in the bathroom of a club can be triggered by dance music or even the sound of a toilet flushing. Odors, like the musty smell of heroin cooking in a spoon prior to injection, are particularly evocative. In his moving autobiography of teenage heroin addiction, *The Basketball Diaries*, Jim Carroll writes of a friend who tried to kick his heroin habit by seeking spiritual solace in the Catholic church of his youth. However, the smell of the church incense reminded him so much of the musty-sweet odor of bubbling heroin that he felt an overwhelming craving and rushed home from Mass to shoot up once again.

As addiction develops and tolerance, dependence, and cravings emerge, the euphoria produced by the drug gradually drains away. Pleasure is replaced by desire; liking becomes wanting. In everyday speech, we may say of an alcoholic, "She really loves to drink," or of a cocaine addict, "He must love to get high." We imagine that drug addicts experience more pleasure from their drug of choice than others and that this motivates their compulsive drug-seeking. However, most active addicts report that they no longer derive much pleasure from their drug of choice. Accumulating evidence indicates that once the trajectory of addiction is under way, pleasure is suppressed, and it is wanting that comes to the fore. Unfortunately, pleasure in the drug itself isn't the only sensation diminished in addicts, for addiction produces a broad change in the pleasure circuit that also affects the enjoyment of other experiences, like sex, food, and exercise.

Drug addiction, whether to cocaine, heroin, alcohol, or nicotine, is notoriously difficult to break. Relapses, even after months or years of drug-free living, are common, and most abstaining addicts have had to make multiple attempts to get clean and stay clean. It is well known that relapse can be triggered not only by sensory cues that are associated with past drug use (like particular people, odors, music, rooms, etc.) but also by emotional or physical stress. A central insight in recent years has been that the later phases of addiction, characterized by cravings and relapse, are associated with strong and persistent memories of the drug-taking experience. Addictive drugs, by co-opting the pleasure circuitry and activating it more strongly than any natural reward, create deeply ingrained memories that are bound up in a network of associations. Later, these memories are strongly activated and linked to emotional centers by drug-associated external cues and internal mental states. If that wasn't enough to battle, addicts who relapse and take even a small dose of a drug after a period of abstinence will get a pleasure rush that's much stronger than that felt by a first-time user, an effect called drug sensitization.

Habitual drug use produces a long-lasting rewiring of the addict's brain, which is manifested at the level of biochemistry, electrical function, and even neuronal structure. If we want to understand and treat addiction at the level of molecules and cells and develop therapies to help people break their addiction and stay drug free, then we need to look for drugs that can produce *persistent* cellular and molecular changes in the brain. Of course, the first place to focus such efforts is in the medial forebrain pleasure circuitry, and the good news is that we don't have to start from zero. Neuroscientists have already worked out some aspects of how memory is stored in the brain through experience-driven cellular and molecular changes, and these insights can be applied to the brain's pleasure circuits and the problem of addiction.

～～～

Oslo, Norway, 1964. A malaise had settled over the community of neurobiologists investigating the biological basis of memory. Because memories obviously could last for the lifetime of an animal, they had expected that experience should produce long-lasting changes in neuronal function to underlie memory traces. Their best guess for the aspect of neuronal function changed by experience was synaptic transmission, the process by which a spike enters an axon terminal and triggers the release of neurotransmitter molecules, which then diffuse across the synaptic cleft to bind receptors and thereby activate the information-receiving "postsynaptic" neuron. Synaptic transmission is the fundamental mode of rapid communication between neurons and is central to information processing in the brain. The dominant hypothesis was that particular patterns of neuronal stimulation delivered to neurons via electrodes (thereby mimicking actual experience in the world) would produce long-lasting changes in the strength of synaptic transmission. The problem at the root of the neurobiologists' malaise was that no evidence whatsoever existed to support this postulated mechanism. The longest-lasting changes that had been recorded persisted for only a minute or two—a time scale that was totally insufficient for memory storage.

In 1964, Terje Lømo was a doctor in the Norwegian navy, soon to be discharged. On leave in Oslo to look for a job, he bumped into the neurophysiologist Per Andersen while walking down the street. After an animated conversation about synapses and neurons, he agreed to join Andersen's laboratory as a Ph.D. student. At that time, recordings of synaptic function in the brain were, for

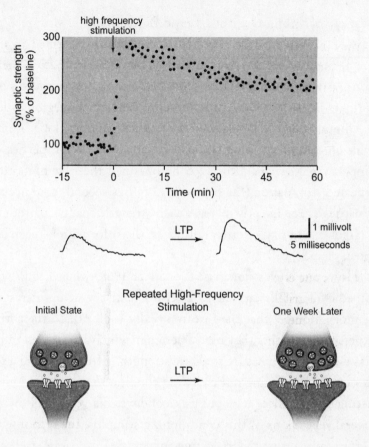

Figure 2.4 Long-term synaptic potentiation (LTP) is a long-lasting, use-dependent increase in the strength of synaptic transmission that can be triggered by particular patterns of activity. The top panel shows a time course in which the amplitude of synaptic strength is plotted on the y-axis and time on the x-axis. LTP is induced by brief high-frequency stimulation (100 pulses, each delivered every 10 milliseconds) at t = 0 min. The middle panel shows electrical traces representing the activation of an excitatory synapse that uses the neurotransmitter glutamate. The amplitude of the deflection in membrane voltage indicates the strength of the synapse, which increases after induction of LTP. The bottom panel shows a schematic diagram of some of the changes that can underlie LTP: increased neurotransmitter release, increased density of neurotransmitter receptors on the postsynaptic neuron, and growth of both the axon terminal and the postsynaptic region, which is called the dendritic spine. Illustration by Joan M. K. Tycko.

a variety of technical reasons, very difficult to make. Most recordings of neuron-to-neuron synapses had been performed in the spinal cord. (These yielded the brief synaptic facilitation mentioned earlier.) Andersen had developed techniques to record synaptic transmission in anesthetized rabbits in a brain region called the hippocampus, buried deep within the temporal lobe. Lømo took up these techniques and began to probe the properties of hippocampal synapses. In 1965 he obtained the first hints that repeated stimulation (120 pulses at 12 pulses/second) could cause synapses to persistently increase their strength, that is, to produce a larger excitatory electrical effect in the information-receiving (postsynaptic) cell.

However, it was not until the fall of 1968, when Lømo was joined by Tim Bliss, a visiting British scientist with an interest in memory storage, that the research really took off. In their first experiment together they used a design in which a single test pulse was delivered to measure synaptic strength. After recording a series of stable baseline responses, a "conditioning stimulus" consisting of 300 pulses at 20 pulses/second was delivered. Following several repetitions of this conditioning stimulus, the response to the test pulse was larger, reflecting an increase in synaptic strength. Most important, this increase persisted not just for a minute or two, but for many hours—as long as the recording could be maintained (Figure 2.4). That day in 1968 marked the first real glimpse of a memory storage mechanism in the brain and began the modern era of memory research in which memory is analyzed at a cellular and molecular level.[18]

Lømo and Bliss called their new phenomenon long-lasting potentiation, or LLP. However, as often happens in science, this name didn't stick, and it is now known as long-term synaptic potentiation, or LTP. (Bliss once quipped that the acronym LLP didn't catch on because it made the speaker seem as if he were in need of urgent assistance.) Starting in the 1970s, LTP created tremendous excitement among memory researchers not only because of its duration, but also due to its relevance to a well-known neurological case.

H.M. was a patient who had undergone surgery to control otherwise intractable epilepsy. The operation, which involved bilateral resection of his hippocampus and some surrounding tissue, cured the epilepsy, but left him with two profound memory impairments: He could no longer recall facts and events from a period of one to two years prior to the surgery, and, even stranger, he could no longer form any *new* memories for facts and events, a phenomenon called anterograde amnesia. These symptoms implicated the hippocampus in memory storage. At that time, the prevailing theory was that LTP and its later-discovered mirror twin, long-term synaptic depression, or LTD, were rare phenomena that would be found only at a few specialized synapses in the brain that had particular roles in memory storage. That has turned out not to be the case at all: LTP and LTD are nearly universal properties of synapses and can be found everywhere from the spinal cord to the most recently evolved portions of the frontal cortex, and almost every brain region in between. Even the most ancient parts of our brain—the parts we share with fish and lizards, regions that control basic functions like spinal reflexes, breathing, temperature control, and the sleep/wake cycle—have LTP and LTD and hence the capacity to be modified by experience. So, it turns out, do our pleasure circuits.

~~~~~

Let's briefly recap what we've learned in this chapter in order to formulate some ideas about drug addiction and lasting changes in brain circuitry. We know that addictive drugs activate the medial forebrain pleasure circuit—in particular, the dopamine neurons of the VTA—and that this activation is central to the euphoria that these drugs produce. We also know that sensory experience can write memories into brain circuitry. These memory traces are formed, at least in part, at synapses, by means of LTP and LTD. Finally, we know that there is a time course of addiction in which initial euphoria becomes overlaid with drug tolerance and dependence and intense cravings that can last for years after drug taking ceases. Continued cravings lead to a high incidence of relapse, often triggered by stress. These facts together suggest an interesting and straightforward general hypothesis: Repeated exposure to addictive drugs produces persistent changes in the function of the medial forebrain pleasure circuit and its targets, including (but not limited to) LTP and LTD, and these long-lasting changes can underlie certain aspects of the trajectory of addiction, in particular, tolerance, dependence, cravings, and relapse.

LTP and LTD are most commonly studied at excitatory synapses that use the neurotransmitter glutamate, like those between the axons of the prefrontal cortex or the amygdala and the dopamine neurons of the VTA. In 1993 C. McNamara and colleagues from Texas A&M University prepared rats in a Skinner box so that lever presses delivered tiny intravenous injections of cocaine. As we've discussed, normal rats will rapidly learn to press the lever at a furious pace to get cocaine (or other addictive drugs). However, the researchers found that when the rats received an injection of a compound called MK-801 that blocks the induction of the most common forms of LTP and LTD before being placed in the Skinner box, they were indifferent to cocaine. They pressed the lever to self-administer the drug only at chance levels.[19] Likewise, when an experimenter manually controlled cocaine infusion, the rats would return to the portion of the cage where the drug was administered because they had formed an associative memory relating that particular location to the pleasure produced by the cocaine hit. However, when rats were pretreated with MK-801, this association was not formed, and the rat continued to explore the cage freely after a cocaine infusion. When MK-801 is administered systemically via an injection in the abdomen, it acts everywhere, not just at the glutamate-using synapses of the VTA. So it is important that a tiny MK-801 injection directly into the brain using a needle that specifically targets the VTA (but not other regions like the nucleus accumbens) can also block cocaine-evoked place preference.

These findings with MK-801 pretreatment suggested that LTP and/or LTD of the excitatory synapses received by the neurons of the VTA was induced in rats as a result of taking cocaine. To test this idea, rats were given a single large dose of cocaine, and following a twenty-four-hour waiting period, the strength of the synaptic connections between the excitatory glutamate-containing axons and the neurons of the VTA was measured. Remarkably, a single cocaine dose produced robust LTP, rendering those synapses stronger. Cocaine-evoked LTP was also prevented by prior treatment with MK-801. Later experiments showed that this LTP was still going strong when the VTA glutamate synapses were measured three months after a single dose.

Other work has shown that drug-evoked LTP in the VTA was not limited to cocaine, but could also be produced by doses of amphetamines, morphine, nicotine, and alcohol. Importantly, non-addictive drugs like the antidepressant fluoxetine or the mood stabilizer carbamazepine did not evoke LTP, demonstrating that LTP in the VTA is not a general effect of all drugs that act in the brain. Another key finding is that these addictive drugs did not produce LTP at synapses throughout the brain, or even all of those that use glutamate as a neurotransmitter; glutamate-using synapses in the hippocampus, for example, are not persistently altered by cocaine or morphine treatment.

So what does the discovery of drug-evoked LTP of glutamate-using synapses in the VTA mean for behavior? Recall that these synapses convey information from the prefrontal cortex, which is important for converting sensory information into plans and judgment, and from the amygdala, which processes emotional information. When these excitatory synapses are made stronger by drug-evoked LTP, subsequent sensory cues and emotions will more easily activate the VTA neurons, causing them to release dopamine in their target regions. One reasonable hypothesis is that drug-evoked LTP in the VTA is necessary for learning the association between the pleasure evoked by the drug and the sensory cues and emotional state that accompanied it.

These recent findings are very exciting in that we now actually have explanations for some aspects of addiction at the levels of cells and molecules in the brain. But we need to put them in perspective: Drug-evoked LTP in the VTA cannot account for the entire biological basis of addiction. After all, this LTP is produced by a single dose of drug—and holds true even for single doses of less dangerous drugs like alcohol or nicotine—which is not sufficient to produce addiction. So, then, what happens to the brain circuitry of rats when they *are* exposed to cocaine repeatedly? Surprisingly, there's no further LTP of the VTA glutamate synapses—a single dose already produces maximum potentiation. However, repeated doses of cocaine do produce an additional form of plasticity that is not seen following a single dose: LTD of inhibitory synapses in the VTA that use the neurotransmitter GABA. Because the action of GABA is opposite to that of glutamate, LTD of GABA-using synapses will reduce inhibition of VTA neurons, leading to their excitation, thereby further turning up the gain on the pleasure circuit and producing even more dopamine release in the VTA target regions (see Figure 2.2). This synergistic effect on the activation of

VTA dopamine neurons (LTP of excitatory synapses combined with LTD of inhibitory synapses) is likely to underlie at least a portion of the drug craving seen in the later stages of addiction.

In habitual drug use the repeated barrage of the nucleus accumbens, the dorsal striatum, and the prefrontal cortex by the VTA dopamine neurons produces changes in these target structures as well. After five days of repeated cocaine administration, the nucleus accumbens undergoes a series of alterations. One of these is an increase in the level of the neurotransmitter dynorphin, one of a class of natural molecules called endorphins that have morphinelike effects. Increased dynorphin release in the nucleus accumbens dampens the electrical activity in this portion of the pleasure circuit (and thereby overrides activity in "upstream" structures like the VTA). Activity in the nucleus accumbens is further suppressed through another mechanism: LTD of the glutamate-using synapses that convey information to the nucleus accumbens from the hippocampus, the prefrontal cortex, and the amygdala. Both of these changes in the nucleus accumbens turn down the gain on the pleasure circuit and are likely to underlie early features of addiction: tolerance and dependence. At this stage, in the absence of more cocaine, the action of the pleasure circuit is chronically suppressed, leading to depression, lethargy, irritability, and the inability to derive pleasure from other activities: the mental symptoms of drug dependence/withdrawal.

When rats are given five days of cocaine treatment and are then kept drug free for days or weeks thereafter, mimicking the conditions faced by an abstaining addict, still further neuronal changes occur. One of the most noticeable is in the fine structure of the main class of neurons in the nucleus accumbens. These are called medium spiny neurons, because their long, branching dendrites (the treelike structures where most synaptic contacts from other

saline

cocaine

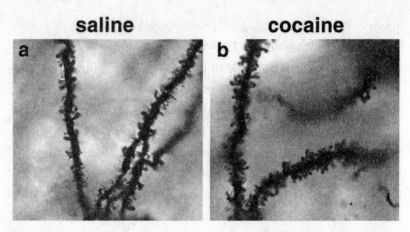

Figure 2.5 Repeated cocaine doses rewire the brain's pleasure circuits. In one study, rats were given either cocaine or saline solution every day for twenty-eight days and then sacrificed two days later. Brain sections were prepared, stained, and observed with a microscope to reveal the spine-covered dendrites of nucleus accumbens neurons. Each tiny nubbin in the images above is a dendritic spine, the point where an excitatory synaptic contact is received. You can see that the dendrites of the cocaine-treated rats appear bushier, showing a higher density of dendritic spines. Reprinted from S. D. Norrholm, J. A. Bibb, E. J. Nestler, C. C. Ouimet, J. R. Taylor, and P. Greengard, "Cocaine-induced proliferation of dendritic spines in nucleus accumbens is dependent on the activity of cyclin-dependent kinase-5," *Neuroscience* 116 (2003): 19–22, with permission from Elsevier.

neurons are received) are covered with tiny nubbins called dendritic spines. The spines aren't just ornamental; they are where the glutamate- and dopamine-using axons from other brain regions form their synapses. In rats that become addicted to cocaine there is an overgrowth of dendritic spines so that these medium-spiny neurons become super-spiny, allowing for increased excitation (Figure 2.5). In addition, each individual synapse received by a medium-spiny neuron undergoes LTP. This LTP doesn't just counteract the LTD seen immediately after five days of cocaine—it overshoots it, leaving the synapse stronger than it was in the predrug state. These two persistent changes in the nucleus accumbens following a period of abstinence have been suggested to underlie drug sensitization, the final hurdle for an addict trying to stay clean.[20]

There's still a lot we don't know about the neurobiology of addiction, but exciting progress has been made that enables us to construct useful models that are subject to experimental testing. We now have a foundation that can be built upon. At present this research is mostly conducted with rats and mice, as we can't yet measure LTP or LTD noninvasively, as would be required in human experiments. (Today's brain scanners aren't up to this task.) Nevertheless, there's great promise for developing therapies to help people break addictions. The first generation of drugs directed against neurotransmitter receptors in the VTA and its target regions are already in use or in clinical trials, and more are on the way. (In chapter 7, "The Future of Pleasure," we'll consider present and future anti-addiction treatments in more detail.) The hope is that these treatments will suppress cravings and prevent sensitization and relapse and help recovering addicts stay drug free. The challenge will be to do so while not compromising other aspects of the pleasure circuit involved in crucial behaviors like eating and sex.

～～～～

These days most of us are willing to believe that drug addiction—including alcoholism—is a disease. Still, we harbor a sneaking suspicion that it's really a disease of the weak willed, the spiritually unfit, or people who are not quite like us. The comedian Mitch Hedberg understood this when he riffed:

> Alcoholism is a disease, but it's the only one you can get
> yelled at for having.

"Goddamn it, Otto, you're an alcoholic!"
"Goddamn it, Otto, you have lupus!"
One of those two doesn't sound right.

Whatever our prejudices, the truth is, given the right circumstances (which can include factors like high stress, early drug exposure or childhood abuse, poor social support, or genetic predisposition), anyone can become a drug addict. Addiction is not just a disease of weak-willed losers. Indeed, many of our most important historical figures have been drug addicts—not only the creative, arty types like Charles Baudelaire (hashish and opium) and Aldous Huxley (alcohol, mescaline, LSD), but also scientists like Sigmund Freud (cocaine) and hard-charging military leaders and heads of state from Alexander the Great (a massive alcoholic) to Prince Otto von Bismarck (who typically drank two bottles of wine with lunch and topped it off with a little morphine in the evening).[21]

From studies comparing identical and fraternal twins it is estimated that 40 to 60 percent of the variation in the risk for addiction is contributed by genetic factors. That said, we are only in the early stages of understanding genetic contributions to addiction. There is no single "addiction gene," and it is likely that a large number of genes are involved in this complex trait.[22] One tantalizing observation concerns the gene for the D2 subtype of dopamine receptor, a crucial component of the pleasure circuit. A particular form of this gene, called the A1 variant, results in reduced expression of D2 dopamine receptors within the nucleus accumbens and the dorsal striatum. Carriers of the A1 variant are, as a result, significantly more likely to become addicted to alcohol, cocaine, or nicotine. Furthermore, among alcoholics, those with the A1 variant tend to be more severely affected, with earlier age of drinking onset, more severe episodes of intoxication, and more unsuccessful attempts to quit. In families with a strong history of alcoholism, brain scanning has revealed that those family members who were not alcoholics had more D2 receptors in the nucleus accumbens and the dorsal striatum than those who were. Taken together, these studies suggest that elevated levels of D2 receptor may be protective against certain forms of drug addiction. Indeed, in rats trained to self-administer alcohol, injection of a genetically engineered virus into the striatum to produce increased D2 receptor expression caused them to reduce their alcohol intake considerably. (A control group, which received an inactive virus, did not show this effect.) We are not at the point of injecting engineered viruses into the brains of human drug addicts anytime soon, but these findings do suggest the D2 receptor as one target for new addiction therapies.

~~~~

If these biological processes are so important in addiction, then where does the social/experiential role come into play in recovery? Do talk therapy or twelve-step groups or prayer or meditation really have a significant function? Given what we know about the genetic predispositions for and biological substrates of addiction, it's easy to conclude that we're all slaves to our genes and our brain chemistry. That's simply not true. The long-lasting changes in neural circuits that are the result of repeated drug use, such as LTP and LTD and structural changes in neurons, can all be produced by one's experience in the world as well. Indeed, these are some of the very mechanisms that enable us to write our experiences into memory and hence confer our individuality. In the brain, causality is a two-way street. Yes, our genes and our neural circuits predispose us to certain behaviors, but our brains are malleable, and we can alter their neural circuits with experience. When an addict goes to talk therapy or engages in mindful meditation to reduce stress or to create associations between drug use and negative life consequences, these actions don't just occur in some airy-fairy nonbiological realm. They create changes in the pleasure circuitry to reverse or otherwise counteract the rewiring produced during addiction. This is the biological basis of social and experiential therapy.[23]

~~~~~

When we say that addiction is a disease, aren't we just letting addicts off the hook for their antisocial choices and behaviors? Not at all. A disease model of addiction holds that the development of addiction is not the addict's responsibility. However, crucially, *recovery* from addiction is. We don't blame someone with heart disease for the development of his condition. Yet once the disease is diagnosed, we do expect him to be responsible for his recovery by eating a healthy diet, exercising regularly, taking his medication, and so on. Similarly, believing that addiction is a disease does not absolve addicts from their responsibility for their own recovery and everything that entails. It's not a free ride.

CHAPTER THREE

FEED ME

There is a war going on for your ass. And no, I don't mean "ass" in a metaphorical sense, like when the funkmaster George Clinton shouts, "Free your mind and your ass will follow." Nor am I using "ass" like Kevin, the kid who had the locker next to mine in seventh-grade gym class, did when he said, "Linden, your mouth just wrote a check your ass can't cash." For the purposes of this discussion, "ass" does not refer to one's corporeal self in a general sense.[1] Rather, I'm speaking literally: There's a war going on for the fat deposits in your *tuchus*, and your brain's pleasure circuits are a major front in that campaign.

~~~~~

In 2008 I ate about 1.2 million calories and loved every one of them. They came in many forms, from fat-laden restaurant meals with pretentious lists of ingredients to sensible portions of vegetarian fare cooked at home to nasty little bags of Cheetos furtively wolfed down in my office with the door closed. I went for many weeks at a time in which I dutifully rode my bicycle for forty minutes every night and whole months where I was an utterly sedentary sofa slug. During that year my weight never fluctuated by more than five pounds and my weight on the first and the last days of that year was identical. Now, putting aside my frustration at this state of affairs, it's remarkable that, with 1.2 million calories of food coming in, my body regulated my appetite and expended precisely the right amount of energy to break even.

My own experience with weight homeostasis is typical for others with unrestricted access to food. In studies where the food intake and energy expenditure of subjects are carefully monitored over a period of weeks to months (which tends to average out day-to-day fluctuations) a remarkable balance between calories consumed and calories burned was observed. When various mammals, from mice to monkeys, are either overfed or starved for a few weeks, their weight soon returns to normal levels when free access to food is resumed. Crucially, our mammalian bodies seem to be able to regulate feeding based on the amount of energy available in the food we consume, not just on the volume of that food. One example of many: When groups of rats were fed nutrient solutions of varying concentrations, they adjusted the volume consumed to achieve a constant inflow of calories. It's a lot like the thermostat in your house: When its thermometer registers a drop in temperature, it sends a signal to the heater to warm the house until the desired set point is reached.

These observations suggest that the brain must receive signals from the body that indicate its weight and that the brain makes use of the signals to modulate appetite and energy expenditure in order to maintain an individual's weight within a fairly narrow range. The signals are received in a structure at the base of the brain called the hypothalamus. The hypothalamus is involved in the control of many basic, subconscious drives and reflexes including sex, feeding, aggression, drinking, and regulation of body temperature.[2] When rats received lesions in a particular subregion of the hypothalamus called the ventromedial area, they became obese. They behaved as if they were starving and compensated with an increase in food intake and a decrease in energy expenditure. Conversely, when a different part of the hypothalamus, called the lateral area, was destroyed, the rats behaved as if they had been overfed. They reduced food intake and increased energy use and thereby became dangerously lean. This is not just a rat trick: These experiments have been replicated in a wide variety of mammals, and humans who sustain damage to the ventromedial hypothalamus (usually from a tumor of the adjacent pituitary gland) will also increase their food intake and become obese.

This model raises one obvious question: How does your hypothalamus know how much you weigh? Let's step back and play God for a moment. If you wanted to build this system, how would you do it? By measuring blood glucose? Fat deposits? Core body temperature? Pressure on the soles of the feet?

This all remained a mystery until 1994, when Jeffrey Friedman and his colleagues at Rockefeller University reported their observations of two strains of mutant mouse, one called *obese* and the other called *db*. (These mutations were not created by scientists using genetic tricks but arose spontaneously in a breeding colony.) Both strains of mice were extremely fat, a trait that was passed on to their offspring in a simple, dominant pattern of inheritance, like eye color. This suggested that obesity in both *obese* and *db* mouse strains resulted from a mutation in a single gene in each case. Friedman's group was able to track down the mutation in the *obese* mice and found that it blocked production of a particular protein hormone, which they named leptin. The leptin protein is only secreted by fat cells. When similar analysis was performed on the *db* mice, it was found that the disrupted *db* gene was respon-

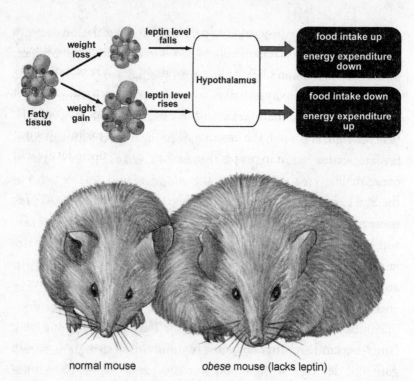

normal mouse
obese mouse (lacks leptin)

Figure 3.1 Leptin is a hormone produced by fat that acts in the brain to reduce appetite and increase energy expenditure, thus keeping weight constant. Top: When weight is lost, fatty tissue is reduced in mass, and so less leptin circulates throughout the body. This triggers increased food intake and reduced energy use, leading to compensatory weight gain. Conversely, when weight is gained, the increase in fat mass causes leptin levels to rise, suppressing feeding and increasing metabolism and activity to burn more energy, resulting in weight loss. Bottom: When the leptin gene is deleted in mice (either through random mutation of the DNA or by genetic engineering), no leptin is produced and severe obesity results. This drawing shows a leptin-lacking *obese* mouse on the right compared with a normal mouse on the left. The same effect is seen in mutant mice that lack the leptin receptor. Illustration by Joan M. K. Tycko.

sible for encoding a protein that functions as a leptin receptor: When it binds circulating leptin at the cell surface, it sets in motion a biochemical cascade inside the cell. Most provocatively, the leptin receptor is expressed strongly on neurons in those areas of the hypothalamus that cause obesity or leanness when destroyed.[3]

So with Friedman's key findings we now have a reasonable hypothesis for how the hypothalamus can sense body weight and use that information to maintain it within a narrow range (Figure 3.1). When weight is gained, the amount of body fat increases, and since fat cells secrete leptin in proportion to their mass, leptin levels will consequently rise. Leptin circulates in the blood and crosses into the brain, where it is sensed by leptin receptors expressed on neurons in the hypothalamus. Activation of those neurons by leptin suppresses appetite and increases energy expenditure. When weight is lost, the system works in the opposite direction: Less fat means reduced levels of circulating leptin, increased appetite, and reduced energy expenditure.

So far, the evidence that supports this hypothesis is quite promising. Leptin levels in the blood do indeed increase with weight gain and decrease with weight loss. Injections of leptin in *obese* mutant mice cause them to reduce food intake and lose weight (and these injections work even if tiny doses are delivered directly to the hypothalamus). Injections of leptin in *db* mutant mice have no effect, because there are no leptin receptors in the hypothalamus for the exogenous leptin to activate.

Of those people who are morbidly obese,[4] less than 1 percent harbor DNA mutations that disrupt the function of the leptin gene — a low rate of incidence that is not surprising, as leptin-deficient humans and mice are both infertile, so these mutations do not pass readily to subsequent generations. However, it's encouraging that leptin-deficient patients can respond to exogenous leptin with a substantial reduction in food intake and subsequent weight loss. I. S. Farooqi and coworkers at Addenbrooke's Hospital in Cambridge, England, reported a case of a nine-year-old leptin-deficient girl with a nearly insatiable appetite. She ate enormous meals and constantly demanded between-meal snacks, as her leptin-deprived brain essentially made her feel as if she were starving. She weighed 208 pounds, which necessitated surgery on her legs to enable her to walk properly. After a year of leptin treatment, she had lost thirty-four pounds, almost all of which was fat. Her food consumption was reduced by 42 percent (which accounts entirely for her weight loss), and she reported that she no longer felt constantly hungry.[5] The small number of patients who suffer from morbid obesity produced by mutation of the leptin receptor unfortunately cannot benefit from leptin therapy, much like *db* mutant mice.

~~~~~

While the leptin homeostatic system explains how the brain can receive information about long-term changes in weight as indicated by body fat, it doesn't account for the short-term regulation of appetite. For example, what signals drive the initiation of eating? It used to be thought that reduced blood glucose levels were the primary trigger for the onset of a meal. However, more recent evidence indicates that eating is biochemically induced only in cases of severe starvation. More typically, in situations where food is abundant, meal onset is driven more by sociocultural and environmental factors.

How is the brain notified about the status of feeding during a meal? You don't actually add fat mass during the course of a meal, so a different, rapid signal is required. In the short term, caloric intake is biochemically regulated by signals of satiety that influence the end of a meal. Sensors in the cells that line the stomach and the intestines can provide information to the brain about both the chemical and mechanical properties of the ingested food. (The chemical properties are things like sugar and protein levels, and the mechanical properties are mainly sensed by how much the gut is stretched.) Signals from the gut are conveyed through the secretion of protein hormones. There are several of these gut hormones that signal the brain in different ways: Some reach the brain directly through the bloodstream, while others activate neurons to send it electrical signals. Let's consider one of these gut hormones as an example. When nutrients from a meal activate cells in the small intestine, a subset of cells lining this region of the gut secrete a hormone called CCK, which binds receptors on the endings of neurons in the nearby vagus nerve and activate it, causing electrical impulses to be passed up into the brain stem to a region called the nucleus tractus solitarius. When this region is activated, it in turn activates the mediobasal area of the hypothalamus, the same region that causes severe obesity if it is destroyed.

The mediobasal hypothalamus turns out to be a critical node in the feeding control circuit. In particular, an even smaller area within the mediobasal hypothalamus, the arcuate nucleus, receives both fast neural signals from the gut–vagus nerve–nucleus tractus solitarius pathway and slow body weight signals from circulating leptin secreted by fat cells (Figure 3.2). The arcuate nucleus contains a mixture of different types of neurons, which produce various effects on feeding when activated. A subset of neurons in the arcuate nucleus that contains the hormone POMC is activated by the nucleus tractus solitarius pathway from the gut and in turn inhibits neurons in the lateral hypothalamus. Activation of the lateral hypothalamus causes secretion of yet another hormone, called orexin, which produces feelings of hunger. At the same time, activation of the POMC neurons in the arcuate nucleus also activates a region called the paraventricular nucleus. Cells in the paraventricular nucleus secrete a hormone called CRH, which produces a feeling of satiety. So when you are eating that stack of pancakes, your gut is gradually sensing both the nutrients released from the food and the stretching of your stomach. These signals are conveyed via the complex pathway we just described to result in inhi-

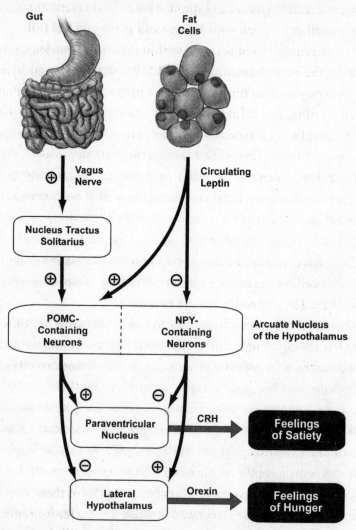

Figure 3.2 The feeding control circuit in the hypothalamus integrates slow body weight signals conveyed by circulating leptin and faster signals from the gut conveyed by the vagus nerve. In the end, the decision to start and stop eating is driven by the competition between two opposing signals: a hunger signal that uses orexin and a satiety signal that uses CRH. One analogy for this is an old-fashioned bathtub with separate hot and cold water taps: The temperature of the bath is determined by the relative flow of hot and cold water. Illustration by Joan M. K. Tycko.

bition of orexin secretion and stimulation of CRH secretion, which work together to block your hunger and make you feel full.

A different group of neurons within the arcuate nucleus, those that use the neurotransmitter called NPY, are unaffected by the vagus nerve–nucleus tractus solitarius pathway, but are inhibited by circulating leptin. Like the POMC-containing neurons, these NPY cells also send axons to both the paraventricular nucleus and the lateral hypothalamus. But their actions are opposite to those of the POMC neurons: The NPY cells inhibit the paraventricular nucleus and excite the lateral hypothalamus. So if you are starving (literally so, not just as you might exclaim while waiting in line for a free table at the House of Pancakes), your fat mass is reduced, and you have less circulating leptin. This means less inhibition of the NPY cells in the arcuate nucleus, resulting in more orexin and less CRH. The end result: You're ravenous.

Why does this circuit have to be so complicated?[6] Why couldn't it be built with a single center that both stimulates hunger and is turned off by a body fat signal or a gut-is-full-of-nutrients signal? Of course, we don't really know the answer to these questions. However, we can speculate that the feeding control system was designed with redundancy and diverging signals to make it more robust and less susceptible to disruption of this crucial behavior. It is also worth noting that feeding is influenced by many factors (time of day, mood, exercise, odors, etc.), and all of these streams of information must be integrated somehow in the feeding control circuit. There are many aspects of this circuit we still don't understand. For example, how and where does orexin act to stimulate and CRH act to suppress appetite?

~~~~~

The idea that eating is primarily a conscious and voluntary behavior is deeply rooted in our culture. We humans are invested in the notion that we have free will in all things. We want to believe that weight can be controlled by volition alone. Why can't that fat guy just eat less and exercise more? He just lacks willpower, right? Not at all. Our homeostatic feeding control circuits make it very hard to lose a lot of weight and keep it off. As weight drops, fat mass decreases and leptin levels decline, triggering the biochemical cascade we just explored, producing signals that both reduce metabolic rate and produce a strong subconscious drive to eat. The more weight that is lost, the stronger the drive to eat will be and the greater the reduction in energy use. This is the sad but unavoidable truth that the multibillion-dollar-a-year diet industry doesn't want you to know.

In the movies we typically see a time-lapse diet montage accompanied by upbeat music: Obese woman on a treadmill, obese woman eating salad, slightly less obese woman on an elliptical trainer, an obviously thinner woman running, ending with lean woman contentedly munching a celery stick and looking self-confident. Very inspirational! However, while moderate weight loss can be maintained through conscious monitoring of food intake and exercise, and dramatic weight loss can be achieved temporarily, it is extremely difficult for most people to maintain an extreme loss of weight over the long term. Even liposuction is only a temporary fix: As with dieting, removal of fat from the body reduces circulating levels of leptin, thereby reducing energy use and increasing appetite.

We share our homeostatic feeding control systems with other mammals. The subconscious drive to eat produced by leptin reduction that we experience is essentially the same as that felt by a mouse or a dog. Although humans are able to overlay these subconscious drives with slightly more cognitive control, at the root, we mammals are all the same. The main factor to appreciate in terms of humans and body weight is that for most of our evolutionary history we have not had access to unlimited calories. Also, for most of our evolutionary history, we belonged to hunter-gatherer societies and burned a lot of energy in everyday tasks. In that context it made sense to have a biological control system that set body weight (and appetite) at an optimum level: Too little of either and you'd be at risk of starvation during the next protracted famine. Too much and your mobility would be compromised. Today, when you try to lose large amounts of weight and keep it off, you are struggling against millions of years of evolutionary selective pressure. But wait. It gets worse. We haven't even started talking about pleasure yet.

~~~~~

As we have seen, the pleasure circuit can be artificially activated, hijacked by drugs, or zapped by implanted electrodes. But is this circuit also activated by naturally pleasurable behaviors like eating? The answer is clearly yes. When a recording electrode is positioned in the VTA of a rat's brain, it reveals a burst of neuronal activity when the rat begins to eat and some degree of continuing activity throughout the meal. Furthermore, when biochemical probes are implanted into the target regions of VTA neurons, eating is found to trigger a surge of released dopamine.

When drugs that flood the brain with dopamine, like cocaine or amphetamines, are given chronically, rats eat less and gain less weight. Loss of appetite is also produced when drugs that mimic dopamine (dopamine receptor agonists) are administered. Conversely, drugs that block dopamine receptors increase appetite and energy intake (total calories consumed) and cause weight gain. Cannabis is another drug that is a well-known appetite stimulant. In fact, it is so effective that it can counteract the dangerous appetite-suppressing effects seen in patients undergoing chemotherapy or who are suffering from AIDS. The brain's own cannabis-like molecules, the endocannabinoids, appear to have a role in feeding behavior. Drugs that block the brain's endocannabinoid receptors suppress appetite and reduce body weight. Similarly, mutant mice that lack cannabis receptors in the brain have low appetites and are lean.

The pleasure circuit is also modified by body weight signals. There are leptin receptors on the dopamine neurons of the VTA, and when circulating leptin binds these receptors, a biochemical cascade is set in motion that inhibits their firing and consequent dopamine release in the VTA target areas. When genetic tricks are used in rats to delete leptin receptors only in the cells of the VTA (but not in other brain regions), the rats eat more and gain weight.[7] This effect was further explored in a study in which leptin-deficient patients were placed in a brain scanner and shown images of food. In these patients the activation of certain VTA target regions (the nucleus accumbens and the caudate nucleus) by food images was similar to that seen in genetically normal patients when they are in the starved state. However, after chronic leptin therapy, the leptin-deficient patients showed normal pleasure circuit activation by food images, coincident with a decrease in appetite. When asked to rate how much they liked each of the food images, the leptin-deficient patients gave lower ratings after leptin treatment.[8] Taken together, these results suggest that when you're trying to sustain a substantial weight loss, the reduction in circulating leptin levels will modulate the pleasure circuit to make food actually seem more appealing.

~~~~~

Because food and addictive drugs activate overlapping pleasure circuits in the brain, there are well-known behavioral interactions between the two that are likely to result from this shared brain wiring. For example, starved rats show enhanced motivation for either addictive drugs like cocaine or amphetamines or direct electrical stimulation of the medial forebrain bundle. (They will press a lever faster and longer to receive these pleasures.)

Does this mean that obesity can be considered as a kind of food addiction? Emmanuel Pothos and his colleagues at Tufts University School of Medicine approached this question by breeding rats for several generations, crossing high weight-gainers to other high weight-gainers and low weight-gainers to other low weight-gainers in order to create obesity-prone and obesity-resistant strains. When allowed free access to standard laboratory rat chow for fifteen weeks, the obesity-prone rats ate significantly more food and gained an average of 22 percent more weight compared with their obesity-resistant cousins. Pothos and others have hypothesized that obesity-prone rats have blunted dopamine signaling in their midbrain pleasure circuits, a condition that causes them to eat more in an attempt to achieve a certain target level of dopamine signaling shared by all rats. Indeed, when dopamine levels in the nucleus accumbens were measured, the obesity-prone rats showed a significant reduction in both baseline dopamine levels and the dopamine surge evoked by electrical stimulation of the VTA. Is this reduced dopamine function something the obesity-prone rats are born with, or do they acquire it as they mature? Studies on obesity-prone rat pups found attenuation of dopamine signaling similar to that found in adults, suggesting the former. These findings support the hypothesis that obesity-prone rats must eat more to achieve the same set point of pleasure achieved by smaller meals in obesity-resistant rats.[9]

While obese rats are interesting for their own sake, how well does their model of heritable obesity carry over to humans? Is there evidence for a genetic component to obesity in humans, or is it all the result of environmental factors? Now, obviously, for a significant fraction of the world's population, environmental concerns are overriding: If you don't have access to sufficient nutrition, you can't become obese. Likewise, many sociocultural factors as well as aspects of an individual's life history also come into play (and we'll get into these in more detail later in this chapter). But why, when given access to unlimited calories, do only some people become obese? Most of our cultural influences (and the diet industry) insist that overeating and obesity result from a failure of willpower. However, the evidence from genetics argues strongly against that idea: Data from adoptions and twin and family lineage studies indicates that about 80 percent of the variation in body weight is determined by genes. That's about the same degree of heritability as a characteristic like height, and much greater than that for other conditions that we now clearly regard as running in families, including breast cancer, schizophrenia, and heart disease.

In a small fraction of cases, obesity results from mutation in a single gene like that encoding leptin or the leptin receptor. It's not surprising that mutations in a number of the other molecules we've discussed in the feeding control circuits, including POMC and the receptors for CRH and MCH, can also cause obesity. However, our best estimates to date are that only about 8 percent of the cases of morbid obesity result from mutations in a single gene. In the vast remainder of the population, interactions of multiple genes with the environment appear to be involved.

Does the genetic component of obesity result in overeating, in normal eating coupled with a slow metabolism, or in both? In most studies, when food intake is monitored closely, it appears as if overeating is by far the dominant factor, with reduced energy expenditure contributing, but to a lesser degree. Yes, obese people eat more and exercise less. But why? A large part of the answer will be found in molecular and cellular analysis of the brain's feeding control and pleasure circuits.[10]

Can we reasonably extrapolate from the aforementioned rat

studies and conclude that a blunted dopamine pleasure circuit might drive compensatory overeating in humans? As is the case with rats, eating in humans is associated with dopamine release in VTA target zones, including the dorsal striatum. The benefit of human subjects is that we can talk to them after they've been in the brain scanner and ask them how they felt. These studies revealed not only that eating was associated with dopamine release, but also that the degree of dopamine release could be used to predict how pleasurable the subject rated the experience of eating. Different foods produced different levels of dopamine release, a finding that correlated with the reported pleasure of eating. (For me, Louisiana hot sausages would peg the dopa-meter.) Also, as hungry subjects continued to eat and became sated, the amount of dopamine released in the dorsal striatum was reduced. Not surprising: The first bites of a meal give the most pleasure when you're hungry.

Again, like rats, humans who take drugs that increase basal dopamine signaling show reduced appetite, as well as reduced caloric intake and weight gain, whereas drugs that reduce dopamine signaling produce the opposite effects. So far, so good. Another important observation is that, on average, dopamine receptor density is reduced in the VTA target regions of obese subjects as compared with those lean subjects (a characteristic that can be measured in a brain scanner). But the key question remains: Do obese individuals show reduced dopaminergic activation of VTA target areas in response to food? Is a blunted pleasure response to food involved in obesity?

A recent study by Eric Stice and coworkers at the University of Oregon placed obese and lean subjects, all young women, in a brain scanner while giving them sips of chocolate milkshake through a plastic tube.[11] Not only is chocolate an unusually good activator of brain pleasure centers, but it's a lot easier to run a flexible straw into the mouth of a head-fixed subject in the brain scanner than, say, proffering a pastrami sandwich or a plate of risotto. The main finding was that the obese subjects showed significantly less activation of the dorsal striatum in response to milkshake sips than did the lean subjects, supporting the blunted pleasure hypothesis.

These milkshake-sipping young women also agreed to DNA testing for a common genetic variant called the TaqIA A1 allele, which results in a reduction in the density of D2-type dopamine receptors in the pleasure circuit. Carriers of the A1 allele showed the greatest reduction in milkshake-evoked dorsal striatum activation. When follow-up examinations were performed a year later, A1 carriers also showed significantly greater weight gain than noncarriers. So do some obese people overeat to try to compensate for low-functioning pleasure circuitry? That's likely to be part of the explanation, but there may be another twist. "If you look at the brain response when people are *about to get* the milkshake, obese individuals show *greater* activation of the reward circuitry, not less," Stice observes. "So, ironically, they expect more reward but seem to experience less."[12] It's a cruel double-edged sword: increased craving coupled with decreased pleasure. In fact, this pattern may be a general problem for many forms of compulsive and addictive behavior, not just overeating. Carriers of the TaqIA A1 allele of the D2 dopamine receptor are not only more likely to be obese, but they are more likely to struggle with drug and alcohol abuse as well as compulsive gambling.

~~~~~

Another front in the war for your body fat is located in the test kitchens and offices of restaurant chains, commercial bakeries, and other "food service" corporations. While body mass index is indeed about 80 percent heritable, it's clear that environment and gene/environment interactions also play a major role in determining an individual's weight. One telling statistic is that the average weight of an adult in the United States has increased by about twenty-six pounds between 1960 and the present.[13] Clearly, this is not due to genetic changes in the population. Rather, it's mostly a result of the concerted efforts of these corporations to produce food and drink, served in large portions, that maximally activate the pleasure circuit and thereby contribute to overeating.

Imagine that we're test kitchen chefs for the Ruby Tuesday restaurant chain or the Nabisco Corporation or PepsiCo (which also owns KFC and Taco Bell). Our goal is to create delicious food that people will crave and will return to buy in large quantities. How do we go about this? How do we create foods that activate the pleasure circuit so strongly that they override the satiety and body weight signals that would normally prevent overeating? Basically, we do this by exploiting the mismatch between the food landscape we humans have evolved to navigate (that of our ancestors) and that of the present.

While our ancestral human diet varied for different groups of our forebears who lived in different habitats, it did have certain common features. It was a diet that was mostly vegetarian, with very little fat (probably about 10 percent of total calories) and very little sugar. Sweet flavors were rarely encountered—they typically occurred in ripe fruit or wild honey—and meat was a rare luxury and was usually quite lean when it could be obtained. For inland peoples, salty flavors were almost unknown. There were few foods with high moisture and oil content that would enable them to be chewed and swallowed quickly. Most important, in many locations intermittent famines were regular occurrences, so when energy-dense foods containing fat and sugar were available, it made sense to gorge on them to establish a body fat reserve for anticipated hard times.[14]

The result of this ancestral diet is that we are hardwired from birth to like certain tastes and smells, most notably those of sugar and fat, but also salt. Both humans and rats show much greater activation of the VTA and dopamine release in VTA target regions when eating energy-dense fatty and sugary foods. Like the difference between "chewed" and injected cocaine, this may reflect the different concentration profiles of glucose (or some other food-related signal) as conveyed to the brain: Large, fast-rising pleasure signals are the most rewarding and most addictive. Interestingly, the combination of fat and sugar is superaddictive, producing a significantly larger jolt to the pleasure circuit than either one given alone. Not only will rats in a Skinner box work hard and perform many lever presses for a sweet, fatty food reward, but those that have already eaten their fill of lab chow will readily consume more food if it is sweet or fatty. (Froot Loops cereal works particularly well.) Then again, we didn't really need an experiment with rats to tell us this: Everyone has had an experience of feeling full at the end of a meal but still having "room for dessert." Our love of salt, meanwhile, remains a bit of a puzzle: Rats won't work for it, but humans crave it, possibly as an adaptation to compensate for salt loss from perspiration.

In our role as corporate test kitchen chefs, we don't need rats, Skinner boxes, brain scanners, or even a knowledge of the scientific literature to figure out how to make foods people crave and overeat; all that is necessary is a collection of different recipes to try out on willing subjects.[15] Even so, developing those recipes is actually a very complex business, for to make craveable foods you can't simply add more salt, fat, and sugar to your existing ones. There is, for example, no single ideal salt concentration for food. We tend to like a lot more salt on our chips and crackers than we do on our meat or in our soup. We like intensely sweet foods better if they are combined with fat. We are also more likely to overeat if foods have a combination of contrasting tastes: Ice cream with chocolate and fruit bits mixed in is more compelling than a smooth, single-flavor ice cream. More Buffalo wings are eaten if they are paired with a contrasting dipping sauce (like cool, fat-laden ranch dressing). Sweet and spicy, fatty and salty, spicy and salty are all combinations that work. Contrasting textures are also highly rewarding: A crispy fried exterior with a soft filling will often be the basis for a craveable food. Flavors and odors from cooking fats also provoke a strong response, perhaps because we have an inordinate number of olfactory receptors devoted to fatty odors.

Another thing that test kitchen chefs have discovered is that people will eat more food if they don't have to work too hard to chew and swallow it. Hence much of the meat served in chain restaurants has been mechanically tenderized and injected with marinade. It dissolves in your mouth quickly and is lubricated for swallowing by high water content. In essence, the factory has done half of your chewing and swallowing for you so you can eat more. And finally, one of the simplest strategies is to take advantage of the fact that people tend to finish what's on their plate (or drink what's in their bottle of soda). Bigger portion sizes are a simple and effective way to promote overeating, defeat the body's appetite control system, and sell more food.

~~~~~

The increase in body weight that's swept the United States in the last forty years, and that is also being seen in other affluent countries, is an enormous public health problem. Increases in weight put people at an elevated risk for a range of health problems, including diabetes, cancer, sleep disturbances, heart disease, and hypertension. The good news is that even modest weight loss—from five to twenty pounds, a level that can be reliably sustained through good eating habits and exercise—has significant health benefits. But what about the morbidly obese, who would need to achieve and sustain a dramatic reduction in weight to experience those benefits? As we have discussed, the brain's homeostatic systems will work against such an effort, as they increase appetite and slow metabolism, making a dramatic weight loss very hard to maintain. Bariatric surgery, in which a portion of the gut is resected, is one option, but because it carries significant risk and expense it is elected in a small fraction of cases. Consequently, there has been a concerted effort by pharmaceutical companies to develop safe and effective anti-obesity drugs to augment diet and exercise.

It's worth noting that there are already unsafe drugs available that will reduce appetite and cause weight loss in a broad spectrum of patients. Drugs like amphetamines, which artificially stimulate the midbrain dopaminergic reward circuit, are very effective but are also highly addictive and have disastrous side effects. Likewise, for many years the drug fenfluramine was prescribed for weight loss, typically combined with a weak amphetamine called phentermine in a formulation sold as "Fen-Phen." Fenfluramine acts by blocking transporters that package the neurotransmitter serotonin into vesicles for later release and by reversing serotonin reuptake across the membrane of the presynaptic terminal to secrete serotonin into the synapse. Tragically, fenfluramine produced heart valve disease in about 20 percent of women and 12 percent of men—a condition that progressed long after these individuals stopped taking the drug. It was withdrawn from the market in 1997 and has been the subject of one of the largest product liability lawsuits of all time, with more than fifty thousand claims and a potential liability for the Wyeth drug company estimated at $14 billion.

At present large numbers of weight loss drugs are in various stages of development, from initial testing in rats and mice to late-stage human clinical trials. Not surprisingly, the potential market for a safe and effective weight loss drug is huge, so considerable commercial impetus is driving research in this area. Candidates include a set of drugs designed to enhance satiety signals from the gut.

For example, the drug SR146131 targets the activation of specific receptors for the gut hormone CCK, thereby promoting a feeling of fullness. Another set of drugs is designed to target feeding control and pleasure circuits in the hypothalamus and medial forebrain. Recall that NPY released from a subset of neurons in the arcuate nucleus of the hypothalamus stimulates appetite (Figure 3.2). Animal studies have shown that drugs that bind and inactivate NPY receptors could block this action and promote weight loss.[16]

Another potentially useful class of weight loss drug is compounds that target receptors for endocannabinoids, the brain's own THC-like molecules. This is a clever strategy inspired by the well-known stoner phenomenon of "the munchies": Since smoking cannabis stimulates appetite, perhaps conversely drugs that block the action of endocannabinoids (at the major neuronal cannabinoid receptor called CB1) would suppress appetite. Indeed, the CB1-blocking drug rimonabant (also known by trade names including Acomplia and Slimona), a compound developed by the drug company Sanofi-Aventis, has been approved for treatment of obesity in over fifty countries. There's no question that rimonabant can produce moderate weight loss. In clinical trials, patients receiving 20 mg/day of the drug for a year lost an average of about sixteen pounds, compared to about four pounds for a group that received a placebo. Other measures of body fat, such as waist circumference and triglyceride levels in the blood, were also reduced.

Unfortunately, its use has also raised some very serious concerns about side effects. Rimonabant has been linked to a significant increase in nausea, major depression, and even suicide in obese patients, a finding that has caused the European Medicines Agency to issue a recommendation to doctors to discontinue its use in European Union countries. In a unanimous vote the fourteen members of the Endocrinologic and Metabolic Drugs Advisory Committee of the Food and Drug Administration turned down a 2007 application from Sanofi-Aventis to allow the use of rimonabant for obesity treatment in the United States. It may ultimately be possible to create CB1-blocking drugs that will decrease appetite without involving such serious side effects, and indeed CB1 drugs being developed by Sanofi-Aventis and other companies have subtle differences in their mode of CB1-receptor blocking that could reduce these symptoms.[17]

Now, at this point you may logically be wondering, "What about leptin? You already told us how it can reduce appetite and cause weight loss in people with an inborn leptin deficiency. Why not just use it on all obese people?" Indeed, the biotechnology company Amgen had this idea in 1995 when it paid $20 million to license the hormone. Unfortunately, when Amgen sponsored a large clinical trial of leptin, few obese participants lost weight.[18] In hindsight perhaps this is not surprising: The vast majority of obese people have high circulating leptin levels because they have a lot of body fat; adding extra, exogenous leptin doesn't help much. The conclusion from these findings: Most obese people are not leptin-deficient, but rather are leptin-resistant. They lack some of the molecular machinery necessary to transduce circulating leptin into reduced appetite and increased energy expenditure. Little is known about the molecular basis for leptin resistance. It could involve changes in downstream feeding control molecules (like NPY or MCH or CRH or orexins) or their receptors. There are even some indications that leptin resistance can result from a failure of leptin to pass from the blood into the brain through a structure of tightly woven cells called the blood-brain barrier. Understanding the molecular and cellular basis of leptin resistance will certainly yield new insights for developing anti-obesity drugs.

~~~~~

Jane has been feeling totally stressed out. She is eighteen years old and lives with three other girls in a small apartment. She and her roommates bicker a lot, and Jane is clearly at the bottom of the social rank. The others push her around, and as a result she tends to avoid them. When she lived by herself, Jane was slim and ate a balanced diet, but since she has been in this pressure cooker of an apartment, she's taken to snacking all day and all night and choosing high-fat foods over more healthy fare. Her weight and waistline have increased significantly. When she went for a checkup, tests revealed high levels of stress hormones in her blood. These days, when one of her domineering roommates approaches, Jane will grimace, emit a submissive squeal, and retreat to the corner.

"Jane" is a macaque monkey living at the Yerkes National Primate Research Center, and I've adapted her story (with some literary license) from a recent report by Mark Wilson and coworkers that appeared in the scientific journal *Physiology & Behavior*.[19] Their findings are remarkable in their parallels with the well-known human phenomenon of stress-induced eating. Not only did Jane and the other low-social-status monkeys in the experiment eat more, but they began to eat "between meals" and they changed the composition of their diet to include more high-fat comfort foods.

Moderate stress will, in fact, stimulate appetite in a wide variety of mammals, from rodents to humans. In several rodent species, chronic restraint stress, forced swimming in cold water, or social stress produced by introduction of a dominant intruder animal triggers increased feeding and the choice of calorically dense foods with high levels of fat and sugar. This causes weight gain and especially the addition of abdominal body fat.

These results suggest that there is some stress-evoked biochemical signal that modifies the feeding and/or pleasure circuits to trigger overeating of comfort foods. Stress triggers a signaling cascade in which neurons of the hypothalamus secrete corticotropin-releasing hormone (CRH), which travels a very short distance through the bloodstream to the adjacent pituitary gland, where it activates pituitary cells to secrete corticotropin itself (also called adreno-corticotropic hormone, or ACTH) into the bloodstream, which then disperses it throughout the body. One important target is the adrenal gland, which when stimulated by corticotropin releases yet another hormone, corticosterone. Corticosterone and its metabolites can pass into the brain, where they contribute to brain responses to stress. The role of corticosterone in feeding is supported by studies showing that corticosterone injections can substitute for stressful experience in triggering overeating. One interesting implication of this work for us humans is that behavioral strategies for stress reduction (like meditation or exercise) can reduce the amplitude of stress hormone surges and are thereby effective in reducing stress-triggered overeating.

While moderate stress can trigger overeating, severe stress can have the opposite effect, suppressing appetite. In humans, for example, it is typical to have a short-term decrease in appetite while grieving the loss of a loved one. This effect of severe stress appears to be conserved across mammalian species as well. When the intensity of either restraint or social stress is increased, rodents will also decrease their food intake.

～～～

Overeating is not the only compulsive behavior that can be triggered by stress. As discussed in chapter 2, stress is often a trigger for the use of certain drugs—such as alcohol, heroin, nicotine, cocaine, and amphetamines—that activate the pleasure circuit. Stress plays a particularly critical role in triggering relapse following a period of abstention in addicts: More than 70 percent of relapsing addicts report a particularly stressful event as precipitating their return to drugs. As with overeating, this finding emphasizes the importance of behavioral stress-reduction techniques in programs to help addicts stay clean. It also suggests that drugs that interfere with stress hormone responses, such as blockers of the CRH receptors, hold promise for the treatment of both stress-induced drug relapse and stress-induced overeating.

How does stress influence the midbrain pleasure circuit (or the feeding control circuits)? The short answer is that we don't really know. However, there are some tantalizing initial clues. Recall that twenty-four hours after a single exposure to cocaine, the excitatory glutamate-using synapses received by VTA dopamine neurons express LTP. This change, which will result in greater dopamine release in VTA target areas, could also be produced by nicotine, morphine, amphetamines, or alcohol. Amazingly, even brief exposure to stress (a rat's five-minute-long forced swim in cold water) also produced LTP of the VTA synapses that was indistinguishable from that evoked by drugs. What's more, the stress-induced LTP could be prevented by pretreatment with a corticosterone receptor blocker. This suggests that drugs and stress rewire the pleasure circuit in overlapping ways and that the stress response to trigger LTP in the VTA requires a stress hormone signaling loop from the brain to the body and back.[20]

In addition to stress actions mediated by corticosterone, there is also evidence that CRH can act directly on synapses in the VTA, as there are CRH-releasing axons that run from the hypothalamus to the VTA. In one recent study by Antonello Bonci and coworkers from the University of California at San Francisco, living brain slices containing the VTA were prepared from both cocaine-treated and saline-treated control mice. (These are kept alive for a few hours in a bath of oxygenated salt solution designed to mimic the milieu of the brain.) When CRH was added to the brain slices of control mice, no change in synaptic strength was produced. But when CRH was added to the brain slices derived from cocaine-treated mice, LTP of the glutamate-using synapses received by the VTA was observed. This is an exciting finding because it suggests a biological framework for how stress could trigger drug relapse.[21]

~~~~~

So we've seen that both certain foods and certain drugs can activate the pleasure circuit. We've also seen that, in many cases, obesity results from food addiction and that food addiction shares many properties and biological substrates with drug addiction, including a strong heritable component and triggering by stress. We know some of the ways in which chronic exposure to some drugs can rewire the pleasure circuit, by changing synaptic structure and function. These findings beg the question: Does chronic exposure to craveable foods—those high in combinations of fat, sugar, and salt—also rewire the pleasure circuit to promote and amplify continued craving?

A recent report by Paul Johnson and Paul Kenny of the Scripps Research Institute suggests that this is indeed the case.[22] They began by allowing a group of rats near-constant access to both standard lab chow pellets and an energy-dense "cafeteria diet" including bacon, chocolate, sausage, cheesecake, and frosting. Forty days later, when compared with a control group of rats fed only lab chow, the cafeteria diet rats showed reduction in the levels of D2 dopamine receptors in the striatum, a central node of the reward circuit. Furthermore, when these groups of rats were implanted with electrodes to directly activate pleasure centers and were allowed to self-stimulate their brains, it was found that stronger electrical pulses delivered to the brain were required to maintain self-stimulation in cafeteria diet rats. Thus it appeared as if many days of the cafeteria diet left the reward circuit partially numbed, an effect that is also seen with chronic cocaine or heroin treatment in both rats and humans.

These findings are interesting and provocative correlations, but they do not speak to the question of whether reduced levels of striatal D2 receptors will actually contribute to continued food craving. To address this issue, the authors took a group of lab chow–fed rats and injected a genetically engineered virus into a subregion of the striatum (the dorsal striatum). This virus was designed to reduce levels of D2 dopamine receptors, and its efficacy for doing so was confirmed with biochemical measurements. Like rats that ate the cafeteria diet for forty days, these rats with artificial reduction of their striatal D2 receptors also showed increased thresholds for brain stimulation reward—they were also partially numbed to pleasure.

These are early days in our understanding of both food and drug addiction, and there's a temptation to overreach. Do these addictions share some common biological and genetic substrates in the brain? Almost certainly. Are those substrates identical? Almost certainly not. Moving forward, it's likely that some of the treatments for drug addiction (behavioral strategies leading to stress reduction and relearning as well as emerging biological treatments for drug addiction that act on the brain's reward circuitry or the stress hormones that modify it) will also be useful in treating food addiction.

CHAPTER FOUR

YOUR SEXY BRAIN

What is your cat thinking while she watches you have sex? Even if you are as sexually conventional as they come in this culture—let's say you're not dressed up in a Dick Cheney rubber mask with clamps on your nipples and Wagner's Ring Cycle playing in the background, or you haven't inserted a Bluetooth-enabled electrical shock probe in your anus that's connected to the Internet to be triggered by wild fluctuations in the Hang Seng stock index, but are in a committed heterosexual relationship, with your partner, in private, in your bedroom, hugging, kissing, petting, licking, having vaginal intercourse—your cat still thinks that you're a freak. And she's right.

What revolts her is the fact that you are engaged in mating during the female's nonfertile time. It likewise makes no sense at all to her that you are sticking with one sexual partner during a given ovulatory cycle. That and this privacy thing—everyone knows that you're supposed to have sex in public, where the social group can watch and join in, if necessary. Finally, if a baby results from this lovemaking, your feline witness will wonder why the male keeps hanging around all the time, helping out or providing resources, the wimp. And don't get her started on that child: Five years old, and it still can't take care of itself?

~~~~~

When I've spoken about our human behavior in regard to drugs or food, I've indicated that we're basically the same as other mammals. While we have a bigger neocortex than a mouse or a monkey, and therefore a greater ability to counteract our subconscious drives with cognitive control, at the root our responses to food and psychoactive drugs are the same as those of our distant mammalian relatives. The same is not true of our mating system. As your cat knows, humans are, relatively speaking, off on the fringe here. In most other mammals, the female advertises her fertility with clear signals: unique sexual gestures and calls, odors, swellings, and so on. Males and females do not typically approach each other for sex outside of these fertile times. By contrast, human females have concealed ovulation, so that there are no obvious displays of a woman's ovulatory cycle. Indeed, while women can train themselves to detect their own ovulation, there is no evidence for an instinctive knowledge of ovulation in humans. One consequence of this is that most human sex, even most human penis-vagina intercourse, is recreational: It's not timed to the ovulatory phase. It even continues in situations when conception is completely impossible, such as during pregnancy or after menopause.

Another way humans are sexually aberrant has to do with our choice of partners. Over 90 percent of mammalian species are highly promiscuous, with both males and females having multiple sexual partners, even on the same day. Humans tend to be monogamous, or at least serially monogamous. Expressed another way, most women have a single sexual partner during the course of a given ovulatory cycle. As a consequence, unlike other mammals, humans have very accurate knowledge of paternity. When genetic population studies are performed, over 90 percent of children are found to be the offspring of the mother's long-term partner or husband. It doesn't matter whether the study is done in Beijing or Chicago or a tiny village in Papua New Guinea, the outcome is the same.

Finally, and perhaps most important, in almost all mammalian species, the male and the female do not form a lasting pair-bond after mating, and so the male has no role in rearing the offspring. In fact, in many cases, the male will even leave the social group. If he does remain, he is unlikely to recognize his own progeny. In humans, lasting pair-bonds are common, and the man typically contributes to the well-being of his offspring (even if it is not through direct care). Of course, in very recent years, changes in social conventions and technology have enabled single motherhood to be a viable human endeavor. But the phenomenon is still quite rare worldwide, and on the evolutionary time scale it has appeared only at the most recent moment.

While humans have a unique combination of sexual behaviors, there are aspects of them that we do share with a few other animals. Bonobos and dolphins, for example, are famous for engaging in recreational sex that is not timed to the ovulatory cycle. Gibbons, prairie voles, and emperor penguins have monogamous mating systems in which the male helps to provide for the offspring. Humans, however, are the only species that displays all of these rare mating behaviors.

Why, then, did we end up this way? The most compelling theory is that the human mating system has been driven by the fact that our species has by far the longest and most helpless childhood of any animal.[1] This is caused by the large volume of the adult human brain in comparison with the size of the mother's pelvis.

The mature human brain, at 1,200 cubic centimeters, would simply not pass through the birth canal. As women know, the newborn brain, which measures about 400 cubic centimeters—roughly the size of an adult chimpanzee's—barely fits, as indicated by the phenomenon of maternal death during childbirth, a uniquely human problem. That newborn with a 400 cc brain will undergo massive postnatal brain development: at a furious pace up until age five and then at a slower pace until about age twenty, when the brain is finally mature. During the period when the human brain is undergoing this extensive postnatal maturation, children are still undergoing cognitive and behavioral maturation. Thus while an orangutan or gray whale mother can raise her offspring successfully without paternal involvement, human single mothers in traditional societies are at a great disadvantage because their children remain so helpless for so long. It is the need to provide care for human children, with their huge, slowly maturing brains, that explains our atypical mating system, with its unusual features of concealed ovulation, mostly recreational sex, monogamy (within an ovarian cycle), and a paternal contribution to child-rearing.

To this point we have concentrated on the generic sexual norm of the dominant human culture, the kind established religions tend to favor: monogamous, heterosexual, procreative. However, it is worth mentioning that there are many common but less broadly sanctioned expressions of human sexuality that are also seen in other species. Masturbation is a frequent practice among a number of mammalian species, including horses, monkeys, dolphins, dogs, goats, and elephants.[2] Both females and males indulge in this pleasure, and, like humans, they have been quite inventive in its pursuit. Males of a number of species, including dogs, goats, monkeys, and guinea pigs, engage in autofellatio, sometimes to the point of ejaculation. There are also a few notable reports of autocunnilingus among primates. A female chimpanzee in captivity has been observed to direct a stream of water from a garden hose at her clitoris. Female orangutans have even employed crude dildos fashioned from tree bark and sticks. In one case, a female porcupine was seen to straddle a stick and then stroll around, causing the stick to vibrate against her genital area. But perhaps the most creative form of animal masturbation is that of the male bottlenose dolphin, which has been observed to wrap a live, wriggling eel around its penis.

In his book *Biological Exuberance: Animal Homosexuality and Natural Diversity*, Bruce Bagemihl of the University of Wisconsin states that homosexual behavior has been well-documented in more than five hundred species and is likely to be present in many more. This includes both male and female homosexual behavior, although the former has been more widely observed. Animal homosexuality is expressed in all of the ways you might imagine and some that probably hadn't occurred to you. Male-male and female-female oral sex has been documented in a number of species, including hyenas and bonobos, as has female genital-to-genital rubbing (Figure 4.1). Male-male anal sex has been seen in sheep, giraffes, and bison, and male bottlenose dolphin pairs will penetrate each other's genital slits. Male Amazon river dolphins will even insert their penises in each other's blowholes in the only known example of nasal sex.[3]

Most but not all homosexual behavior seen in animals would be more accurately described as bisexuality. In many species heterosexual contact occurs only during the female's fertile phase, but homosexual behavior is common at other times. In some, like bonobos, it seems as if homosexual behavior, in addition to providing sexual pleasure, also fulfills a social role by diffusing tension and promoting social bonding at the expense of aggression. At present, there are only a few examples of animals engaging in lifelong, purely homosexual behavior, and these have occurred mostly

Figure 4.1 Adult bonobo females engaging in genital-genital rubbing, a common expression of bonobo sexuality that can result in orgasm. Illustration by Joan M. K. Tycko.

among males and mostly in captivity. Nonetheless, in a number of zoos around the world, male penguins of several species have been observed to form stable monogamous couples. They build nests together and use a rock as a surrogate egg. In one well-publicized case, a pair of male chinstrap penguins at the Central Park Zoo in New York City were provided with a fertilized egg, from which they successfully hatched a chick.[4] Likewise, about 6 percent of domesticated rams court and mount other males exclusively, even when estrous females are present.

We can't close out this discussion of animal sexuality without mentioning some even more exotic phenomena. Cross-species sex is most commonly observed in animals in captivity, but examples have been documented in wild populations as well. For example, male moose have been known to have sex with female horses. In a zoo in Siberia, a tiger and a lion were encouraged to mate and bore (infertile) offspring. Genetic analysis has been employed to identify offspring from cross-species hybrids in the wild, providing evidence for sexual encounters between grizzly bears and polar bears.

~~~~~

The Natuurmuseum in Rotterdam, the Netherlands, has a mirrored glass façade that often sustains bird strikes. One day in June 1995, Dr. Cees Moeiliker was in his office when he witnessed the fatal collision of a male mallard duck with his window. When he went to investigate, he found that another male mallard had arrived, which proceeded to rape the corpse continuously for a period of seventy-five minutes. When Moeiliker went to write up these findings for a scientific journal, he discovered that heterosexual mallard necrophilia had already been described in the literature.[5]

How should we think about this disturbing behavior? The most parsimonious explanation is that many animals (particularly males) are sexual opportunists and will attempt sexual contact with just about any species, live or dead. It's possible that, like some humans, some animals are specifically aroused by sex with corpses or other species, but there is no evidence to support this idea.

So our conclusion in light of all these findings is a bit counterintuitive. It's not really kinky or forbidden behavior that makes humans sexually unique—those things are well-represented in our mammalian kin. Rather, it's our most conventional and socially sanctioned mating behavior that is totally aberrant in comparison to other animals.

~~~~~

What about the emotional component of all this physical activity? What happens in our brains when we fall in love? For that matter, what happens to scientists who study the act of falling in love? There's something about this topic that tends to make otherwise hard-nosed biologists and anthropologists get all mushy and literary and start quoting the impassioned lines of Shakespeare, Ovid, and Dante in their scientific papers. In this spirit I would like to offer my own all-time favorite love poem. In my view, it gets to the heart of the matter and does so quickly:

> I don't want a physical relationship.
> I just want someone to fuck with my mind.
>
> —personal ad in the *L.A. Weekly* (circa 1979)

~~~~~

Intense romantic love is not just a modern idea, but has been expressed in some of the oldest surviving writings from China, Egypt, Greece, and Sumeria. But is it truly a cross-cultural universal phenomenon, or is it only true of a handful of cultures? To address this question, Helen Fisher, an anthropologist from Rutgers University, surveyed data collected by cultural anthropologists from 166 different societies and found evidence of romantic love in 147 of them. (In the remainder, there was no clear indication that romantic love did *not* exist; rather, these were just cases where the anthropologists hadn't investigated the topic.) The description of the mental and physiological aspects of intense romantic love were surprisingly similar across cultures: intense, giddy pleasure, suppression of appetite, distortions of judgment about the beloved (accentuating good traits and diminishing bad ones) and the world ("we've been so discreet—no one knows about us"), obsession, and sexual desire. There's a crucial feedback loop in new love: Not only do we see all the wonderful things in our beloved, but when we look into their eyes we see those same positive feelings mirrored back. When we're in love, in other words, we like ourselves better as well. Finally, in the throes of intense romance, changes in mood become amplified: The highs are higher, and if anything goes wrong (or the love is unrequited) the lows are lower.[6]

How do these characteristics of intense romantic love correspond with brain function? To address this question, Lucy Brown, a neurobiologist from Albert Einstein College of Medicine, and her colleagues recruited men and women in the early stages of a relationship (average seven months) who reported being "madly, deeply and passionately" in love.[7] They were imaged in a brain scanner while looking at a photo of their beloved's face. As a control, and after a distraction task to let their ardor cool, the subjects also looked at a photo of an emotionally neutral acquaintance, matched for sex and age—the logic here being that those brain regions that were activated (or deactivated) by the former but not the latter represent the neural substrate of romantic love, as opposed to simple recognition of familiar faces. Of course, this type of study is merely correlational: It does not prove that the particular activated or deactivated regions actually underlie the feeling of being in love. We must also question to what degree the results are vision-specific: What would the brain's activity look like, for example, in response to the beloved's voice or scent? Nonetheless, the pattern of brain changes specific to viewing the beloved's face was remarkably consistent with the lovers' self-reports.

The intense, euphoric pleasure that comes with falling in love? That corresponds to strong activation of the dopaminergic pleasure circuit—the VTA and its targets, like the caudate nucleus. As we have discussed, this pattern of activation is similar to responses to cocaine or heroin.[8] Distortions of critical faculties related to the beloved? These might result from a deactivation of the prefrontal cortex, a judgment center, as well as deactivation of the temporal poles and parietotemporal junction, cortical regions involved in social cognition. Deactivation of certain portions of the prefrontal cortex is also found in obsessive/compulsive disorder, which indeed shares some aspects with new love. While the sample sizes were small, no significant differences were seen between men and women in this study, and the sexual orientation of the subjects was not reported. (In future work, it might be of interest to explicitly compare men and women, gay, straight, and bi.) More recently, Brown's team has extended these findings in some interesting ways. As a first step toward addressing cross-cultural relevance, the experiments were repeated using a similar population of young men and women in Beijing, China. The results were identical.

Social psychologists who have interviewed people in long-term relationships find that the intense, initial phase of romantic love typically lasts from nine months to two years, to be replaced, in most couples, by a less intense form of loving companionship. Given what we know about the distortions of thought and self-image and the sexual obsession that accompany the early phase of love, one has to wonder about our laws. Most states in the United States require a six- to twenty-four-month delay before granting a divorce, but anyone can get married immediately. One could make a case that to promote good, long-term marriages, the delay should be mandated on the front end.

A small number of people do report that their feelings for their partner are just as intense ten or twenty years on as they were soon after they first met. Most of these individuals seem to be telling the truth about the depth of their emotions. When Brown's research group performed the lover's-face brain imaging experiment on subjects whose love relationship had lasted for ten years or more, they found an interesting result. Most of the long-term lovers no longer showed strong activation of the VTA dopamine center—the other brain changes were mostly intact, but the pleasure circuit no longer got that cocaine-like jolt. However, in the small group that still reported being intensely in love, the VTA pleasure circuit remained strongly activated by the image of the lover's face. That interesting result validates the idea that a minority of couples can indeed keep that glow of new love burning beyond the initial infatuated state of a relationship. However, the ultimate causality here is not clear. Are there some people who can sustain intense love because they were born that way, or is there something about particularly well-matched couples that keeps the dopamine flame lit?[9]

~~~~~

Are the brain systems activated in intense romantic love simply the same as those that are driven by sexual arousal? In interpreting the brain activation of new lovers, are we just putting a soft-focus amorous gloss on mere carnality? Our own experience suggests that these drives are dissociable: One can be sexually aroused without being in love, and one can be in love without being sexually aroused. This begs the question: How do the patterns of brain activation produced by sexual arousal compare with those in the new lover's face task?

In recent years there has been a series of studies in which subjects have undergone brain scanning while viewing still photos or videos with various types of sexual content. One particularly well-designed study was performed by Kim Wallen and colleagues at Emory University. They started by collecting a set of explicit images of female-male couples engaged in heterosexual acts that were ranked as sexually arousing in questionnaires given to their pool of fourteen male and fourteen female heterosexual young adult subjects. Their subject pool was chosen to exclude people who did not rate these images as arousing (about 16 percent of women and none of the men).[10] In the end, they had a pool of subjects who rated the test images as equally arousing. As is typical in these experiments, control images were used for comparison: photographs of couples interacting in nonsexual ways. When viewing couple-sex images, both men and women showed strong activation of key elements of the pleasure circuit, such as the VTA and the nucleus accumbens/ventral striatum. Thus both the intense, romantic lovers in the beloved's face task and our smut-viewing subjects here showed substantial pleasure circuit activation. However, the sexual images did not evoke deactivation of the judgment and social cognition centers seen in the lover's face task. Instead, activation of a broad range of cortical regions was produced, including areas implicated in visual processing, attention, motor, and somatosensory function. In a way, this confirms what we already knew from personal experience: While falling in love and feeling sexually aroused share some overlapping pleasurable feelings, they are also quite distinct.

While the couple-sex images evoked a common set of brain region activations in men and women, they also elicited some sex-specific effects. In particular, men showed strong activation of the hypothalamus and the amygdala, an emotion center. These activations were strongest for those particular images that were rated as most arousing. The hypothalamus contains tiny subregions that are crucial for triggering male sexual behaviors, but the resolution of the brain scanner was not sufficient to determine if it was those particular regions that were activated in the men in this study. Do such male-female differences in brain activation reflect sociocultural influences, or genetic or epigenetic underpinnings, or both? Unfortunately, brain scanning studies cannot resolve this issue.

～～～～

For both men and women viewing sexual stimuli, whether still photos or video, there is a good correlation between self-reported sexual arousal and activation of the medial forebrain pleasure circuit. When neutral stimuli, like sports scenes or landscapes, were interspersed among the sex scenes, pleasure circuit activation subsided, and then revved up again when the sex images were resumed. Paul Reber and his colleagues at Northwestern University expanded upon these findings in an interesting way by designing a study in which twelve homosexual and twelve heterosexual men underwent brain scanning while observing explicit images of either male-male or female-female sex. In this case, the control images consisted of males and females playing sports.[11] They found that both gay and straight men exhibited category-specific brain activation that matched their sexual orientation and their self-reported sexual arousal: Gay men were activated by male-male sex, and straight men by female-female sex. Now, you may be wondering, why didn't they use images of heterosexual activity as an optimal stimulus for the straight men? The reasoning behind this was that some straight men are put off by images of men having sex, even with women, and that some gay men are aroused by images of men having sex, even with women, so responses to the female-female and male-male sex images are more easily interpreted. Similar results of category-specific activation of the medial forebrain pleasure circuit in gay and straight men and women were found in a study that used an even simpler stimulus: photos of male and female genitalia in the excited state.[12]

So we have established that, for both men and women, homosexuals and heterosexuals, there is a good match between brain activation (particularly of the medial forebrain pleasure circuit) and self-reported sexual arousal in response to visual stimuli. How do these two correlated measures match up with genital responses? For the men, determining that is a fairly straightforward procedure, as penile erection can be measured using a technique called circumferential penile plethysmography. This involves a condom-like device outfitted with a strain gauge to measure changes in the girth of the penis. When Bruce Arnow and coworkers at Stanford University performed an experiment in which heterosexual men were presented with video snippets of explicit heterosexual activity, interspersed with scenes of sports and landscapes, there was agreement across all three measures: The sex video produced strong brain activation of pleasure centers (and some other regions) and penile erection, and the men reported feeling sexually aroused. Landscapes and sports did not evoke significant changes in any of these measures.[13]

There have also been a number of studies measuring penile erection without simultaneous brain scanning that have further explored the range of subjects and stimuli. Meredith Chivers, Michael Bailey, and their coworkers at Northwestern University and the Centre of Addiction and Mental Health in Toronto compared the genital responses of heterosexual and homosexual men (and also women, whom we'll discuss in a moment) to a series of video clips depicting male-male, female-female, and female-male sex, animal copulation (bonobos or chimps), male nude exercise, female nude exercise, female masturbation, male masturbation, and control nonsexual images, all in a random order.[14] The results were fairly straightforward: Men's erections and their reported sexual arousal were well-matched. Gay men were most aroused by male-male sex images and straight men by female-female sex images. Images of heterosexual activity produced an intermediate level of both subjective and genital arousal for both straight and gay men. Likewise, in both straight and gay men there was an increasing subjective and genital response to sexual images, with nude exercise being lowest, masturbation next, and partner sex being most arousing. In both gay and straight men, the brain activation, subjective arousal, and genital response all seem to match up and are largely category-specific to images of the sexual partner of choice.

What about bisexual men, who in a recent national survey

constituted about 1 percent of the male population? We might expect them to show genital responses to both male and female sexual stimuli. If this hypothesis is correct, men with strong bisexual feelings need not have exactly the same degree of genital response to male and female sexual stimuli. But one would expect that, on average, their arousal to male stimuli should exceed that of heterosexual men and their arousal to female stimuli should exceed that of homosexual men. When the bisexual men in this study were queried about their subjective feelings of arousal, they reported similar responses to male and female images. However, the genital responses did not correspond: Most of the bisexual men in the sample had erections in response to male but not female stimuli, in a pattern reminiscent of homosexual men. A few of the bisexual men had responses to female but not male stimuli, similar to heterosexual men. But significantly, and counter to our hypothesis stated above, there was no distinct pattern of genital response to sexual stimuli in bisexual men.[15] What does this mean? It's not entirely clear. It may be that genital responses to female images in the majority of bisexual men are present in real life, but are extinguished by the decidedly unsexy environs of the lab. Or it may be that these measures do reflect real-world genital responses and that male bisexuality, while clearly a genuine phenomenon, has a more cognitive than instinctive origin.

~~~~~~

Most men, particularly young men, have had numerous experiences with embarrassing, inappropriate erections. One slang dictionary defines such a "bone of contention" as "the argument-provoking erection you get when watching Olympic beach volleyball on television together with your wife." After one similar experience many years ago, I remarked to my girlfriend that if life were fair, women would be outfitted with vaginal sensors and display lights so that their level of genital arousal would also be public information. "Dream on," she said.

Measuring a woman's genital response is not as straightforward a process. The most common method involves inserting a tampon-sized probe called a vaginal photoplethysmograph. This device, which is connected by a wire to external recording equipment, shines light onto the vaginal walls and measures the color of the light that's reflected as an indicator of blood pooling in the vessels that run there. This accumulation of blood (called vasocongestion) is a precursor to vaginal lubrication, a process that uses exuded blood plasma as a base to form the vaginal lubricant. Vaginal photoplethysmography does appear to measure a specifically sexual reaction: Strong responses are seen to sexual stimuli and essentially no responses to stimuli without sexual content.

So what happened in Chivers and Bailey's experiments with the female subjects? Like men, the women, both heterosexual and homosexual, reported the least subjective arousal in response to the nude exercise images and the most to partner sex, with images of masturbation falling in between. However, that's where the similarity ended. Both heterosexual and homosexual women reported some degree of category-specific subjective arousal, although heterosexual women reported arousal to a somewhat broader range of stimuli than either men or homosexual women. However, this did not correspond well with the genital responses. Heterosexual women showed strong genital responses to images of both females and males masturbating as well as to images of both female-female and male-male sex (as well as heterosexual activity). Homosexual women showed a slight bias in genital response to sexual images involving women, but they still responded robustly to images of male-male sex and male-female sex. In addition, both homosexual and heterosexual women responded genitally to images of (heterosexual) bonobo copulation, while men did not. The emerging conclusion, which builds upon work from several independent laboratories, is that women, both heterosexual and homosexual, feel aroused by a broader range of stimuli than men, and that their genital arousal is triggered by a significantly broader range of stimuli than their reported feelings of arousal.

Why do women's vaginas get wet to such a broad range of sexual stimuli, even stimuli that they report as not being arousing? While we don't yet have a data set in which genital, brain scanning, and subjective arousal responses are measured simultaneously in women, the indications so far are that the subjective responses and the brain scanning data in women correlate well, but that the vaginal measure does not. There are several possible explanations for this disparity. Perhaps women's thoughts while watching these sexual images are significantly different from men's, which could in turn influence both brain scanner responses and, indirectly, vaginal responses. Because most women will not consent to take part in a study that involves vaginal and/or rectal probes and sex videos, it may be that these studies are measuring a group of women who are not entirely representative of the broader population. Yet another possibility, voiced by Ellen Laan, Meredith Chivers, and others, is that women have an automatic, reflexive vaginal response to a wide variety of sexual stimuli. An evolutionary explanation for this phenomenon is that reflexive vaginal lubrication was adaptive for ancestral women because it reduced the chance of injury or infection in sexual contact, particularly if that sexual contact was rapid and/or nonconsensual. Indeed, surveys of female rape victims have revealed that many women produce vaginal lubrication even during nonconsensual sex, when they report that they were terrified and not aroused in the least.[16] Similarly, Ellen Laan and colleagues have shown that heterosexual women who are undergoing treatment for sexual arousal disorder show vaginal responses to sexual images even though they report little or no subjective arousal.[17] While we don't yet understand the basis of the broadly tuned female genital response to sexual stimuli, we do know that one interpretation is not correct: A woman's vaginal response does not reveal some "truth" about her feelings that contradicts her expressed ideas.

～～～～

If you're standing in the bookstore, flipping through the pages of this book (or if you're clicking through the sample pages on Google Books) to get to the section on orgasm, don't feel embarrassed: You're probably not alone. The first thing to say about orgasm may be obvious, but it bears repeating: Orgasm occurs in the brain, not the crotch. While the most common and reliable means of achieving orgasm is clitoral stimulation in women and penile stimulation in men, some men and women can achieve orgasm solely from stimulation of the mouth, nipples, anus, ears, or other areas. In fact, a small number of women and men can achieve orgasm reliably through thought alone, without the involvement of any touching whatsoever. In a way, this is not completely surprising, given the well-known phenomenon of orgasm occurring during dreaming sleep. In the neurological literature it is well known that some women and men with complete spinal cord transections can achieve orgasm and that epileptic seizure can trigger orgasm as well.[18]

The overall physiological profile of orgasm is remarkably similar in women and men. In both sexes orgasm involves an increase in blood pressure, a rising heart rate, involuntary muscle contraction (including the muscles of the rectum), and an intensely pleasurable sensation. Orgasm is most often accompanied by contraction of the muscles in the wall of the urethra and two pelvic muscles, the bulbocavernosus and the ischiocavernosus, resulting in the ejaculation of semen in men and, in some cases, glandular fluids in women. Women's orgasms, as measured by a rectal probe (Figure 4.2), are, on average, somewhat longer than men's, lasting about twenty-five seconds as compared with fifteen seconds.[19] And of course, many women, but very few men, can experience repeated orgasms with a very short duration between them.

Brain scanning during orgasm is a technically challenging endeavor. In Gert Holstege's lab at the University of Groningen Medical Center in the Netherlands, a female subject's head is fixed tightly in place with an adhesive band within the giant metal doughnut that is the brain scanner. She's hooked up to an intravenous line to deliver a pulse of a radioactive water tracer, which is required for this particular form of brain scanning (called positron

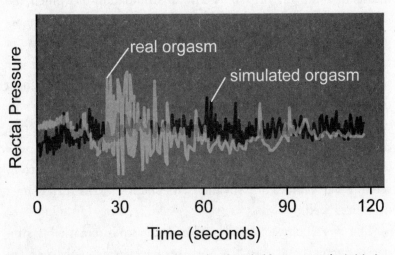

Figure 4.2 Traces from a rectal pressure probe reliably measure the initiation and duration of orgasm. Here, the black trace is the rectal pressure record from a woman who was instructed to simulate orgasm, and the gray trace is from that same woman experiencing a genuine orgasm as a control measure. Adapted from J. R. Georgiadis, R. Kortekaas, R. Kuipers, A. Nieuwenburg, J. Pruim, A. A. Reinders, and G. Holstege, "Regional cerebral blood flow changes associated with clitorally induced orgasm in healthy women," *European Journal of Neuroscience* 24 (2006): 3305–16, with permission from Wiley-Blackwell.

emission tomography, or PET), and a pressure-transducing probe is inserted to measure the rectal contractions that accompany orgasm. She's instructed to close her eyes and to keep as still as possible to minimize brain activation unrelated to orgasm. The subject's male partner begins manual stimulation of her clitoris, and when it appears as if she is approaching orgasm, the tracer is injected through the intravenous line in the hope that the orgasm will occur in the next sixty seconds or so, before the tracer decays. (I find it amazing that *anyone* can achieve orgasm under these conditions.)[20]

When these studies were carried out with heterosexual subjects, the pattern of brain activation seen during orgasm was mostly similar for men and women.[21] Not surprisingly, given that both sexes experience orgasm as intensely pleasurable, it involved strong activation of the medial forebrain dopamine-using pleasure circuit—the VTA and some of its targets, like the dorsal striatum and the nucleus accumbens (Figure 4.3). The cerebellar deep nuclei, a motor control and motor learning center, were also activated and were most strongly activated by those orgasms that produced the strongest rectal contractions. The cerebellar deep nuclei are part of a circuit that computes "motor error." A motor error signal is the difference between a plan for movement and ongoing sensory feedback about how that movement is progressing. When a significant disparity develops between the plan and the sensory feedback monitoring the ongoing motion, that's a motor error. In a way, orgasm, in which you feel yourself sliding over the edge of a waterfall and body movements become uncontrollable, is the ultimate motor error, so orgasmic cerebellar activation makes sense. Deactivations of the left ventromedial and orbitofrontal cortex, judgment and social reasoning centers, were also seen, which likewise makes sense: During orgasm, logical evaluation and reasoning are suspended.

These similarities in the pattern of brain activity during orgasm validate previous work in which twenty-four male and twenty-four female subjects were asked to write descriptions of their orgasms, which were then edited to remove gender-identifying information. When these accounts were given to a panel of seventy psychologists, gynecologists, and medical students, all were unable to distinguish male from female accounts of orgasms.[22]

The only significant sex difference in orgasm was found in an ancient brain-stem region called the periaqueductal gray matter (PAG), which was activated in men but not women. The significance of this activation is not entirely clear. The PAG is known to be activated by painful stimuli resulting in endorphin release from this region, so it's possible that there is a PAG-based endorphin component that's specific to male orgasmic pleasure.

~~~~~

It's easy to begin to conclude that orgasm is just another pleasure buzz, perhaps weaker than heroin but stronger than food. That would be an oversimplification. Orgasm is a multifaceted experience with dissociable sensory and affective/emotional/rewarding components. It's fiery and transcendent and unique. In orgasm, as in all sensory experience, we feel things as an integrated whole. While in everyday life it's not easy to break down such experience into its component parts, in the neurology clinic we have the opportunity to take a look under the hood. Brain stimulation studies have shown that there are discrete locations in the thalamus where stimulation with an electrode can induce orgasms that have all the standard physiological hallmarks (heart rate increases, muscle contractions, etc.) but that just don't feel pleasurable. In essence, these results are the opposite of those of the pleasure circuit stimulation studies we discussed in chapter 1, in which intense pleasure was felt but was not always sexual in nature. Another way that orgasms without pleasure can occur is in epileptic seizures.

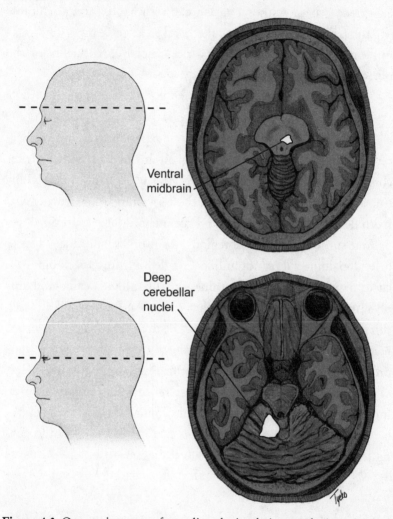

Figure 4.3 Orgasm in women from clitoral stimulation results in activation of portions of the medial forebrain pleasure circuit (top) and the deep cerebellar nuclei, a motor control and motor learning center (bottom). Adapted from J. R. Georgiadis, R. Kortekaas, R. Kuipers, A. Nieuwenburg, J. Pruim, A. A. Reinders, and G. Holstege, "Regional cerebral blood flow changes associated with clitorally induced orgasm in healthy women," *European Journal of Neuroscience* 24 (2006): 3305–16, with permission from Wiley-Blackwell.

While most seizure-triggered orgasms are pleasurable, a subset of patients with seizures involving the parietal and temporal lobes experience them without pleasure. Presumably, nonpleasurable orgasms, however they are evoked, fail to activate the VTA dopamine neurons and their targets, the crucial components of the pleasure circuit.

~~~~~

The poet Jim Carroll described his first experience mainlining heroin as feeling "like 50,000 orgasms all at once." Another writer compared intravenous cocaine injection to "orgasm × 1,000." Putting aside the fanciful arithmetic, there's a strand of neurochemical truth here: Drugs like heroin and cocaine produce a sustained dopamine surge in VTA target regions, while orgasm produces a fairly brief dopamine surge (not unlike a single hit from a crack pipe). As such, drugs that boost dopamine signaling, like cocaine and amphetamines (and even the L-dopa used to treat Parkinson's disease), can prolong and enhance orgasm, while drugs that block dopamine receptors or dopamine release, like some antipsychotic drugs, suppress it. It's worthwhile to note that the positive and negative effects of dopamine drugs on orgasm are also reflected in similar effects on libido and genital responses to sexual stimuli. Cocaine, for example, is one of the few psychoactive drugs that increases libido and promotes penile erection or vaginal lubrication.[23]

~~~~~

Is sex addiction real, or is it just something made up by philandering celebrities to justify their cheating? We've seen from brain imaging studies that both new love and orgasm strongly activate the dopamine-using medial forebrain pleasure circuit. Furthermore, drugs that modulate dopamine signaling in the brain can regulate libido and orgasm. We also know that those drugs that carry a risk

of addiction activate the dopamin___
repeated use of those drugs can mo
of the pleasure circuit in ways that)rtunately short-lived pleasure of
tory of addiction: tolerance, withα, lingering post-orgasmic after-
if repeated orgasms or new love vght to be crucial for sexual pair-
should produce similar changes iniated, in both men and women, by
don't yet know if this is actually th from the pituitary gland, which
believe that it is.   lamus. Crucially, treatments that

In a way, acceptance of the contion of oxytocin on its receptors
of the same problems as acceptinut do interfere with its afterglow.
obesity result from food addiction.cin-releasing system also appears
a necessary activity. Similarly, altspects of social bonding. Oxyto-
almost everyone has sex, and of cns at birth and during breastfeed-
is the traditional way to propagater in developing a mother's bond
mal behavior that almost all of us)n evolutionary theme: It's more
Is the horny teenager who masturbiochemical system for an addi-
dict? The woman who never leavan entirely new one.
night without a new sexual partnerh effectively introduce oxytocin
who seeks out prostitutes every timn developed to help breastfeed-
sex addiction is not always straightime engaging the "milk letdown
criteria are really no different thaleased during nursing. However,
food addiction: been put to other uses. Experi-
                              kers at the University of Zürich

- When the behavior continu ised oxytocin nasal spray before
  ative life consequences to t sting of strangers in a coopera-
  Susan Cheever has observe ith those who received a placebo
  driving of sex addiction.") in group continued to be more
                              betrayed" by other players in the
- When the addict's behavior accompanied by deactivation of
  to "feel normal" and to be evealed by brain scanning. Con-
  stresses of life. ("If I don't ha effect of oxytocin on trust was
  every night, I start to feel an the subjects' readiness to bear

risks, but rather to their increased willingness to accept *social* risks arising through interpersonal interactions. Oxytocin appears to have a complex role in social cognition and behavior that goes beyond mere "trust." One fascinating study showed that oxytocin administration improved the ability of subjects to correctly infer another's emotional state when provided with a photo showing only the eyes.[25]

These observations suggest that oxytocin nasal spray might be useful in treating people with disorders of social interaction and cognition. Recently Markus Heinrichs of the University of Zürich has reported that use of oxytocin nasal spray can reduce fear and stress and thereby enhance interaction in persons with intense social phobias. Initial studies have reported similar salutary effects in people with borderline personality disorder (who are overly suspicious of others).[26] Not surprisingly people have begun peddling oxytocin spray (or what they claim is oxytocin spray) on the Internet. One brand is called Liquid Trust.

~~~~~

What is virtue but the Trade Unionism of the married?

—George Bernard Shaw

Why are some of us true to our long-term partners while others just can't resist the appeal of a new sexual encounter? We don't know the answer to this complex question, but in recent years some interesting ideas have emerged from the study of wild rodents. Voles, also known as field mice, are found around the world in a number of temperate habitats. Some species of vole, like the prairie and pine vole, live in social groups and have a monogamous mating system. After a pair mates, neither the male nor the female is likely to mate with any other vole. In fact, if one partner dies, the remaining partner will typically never mate again. As is common with monogamous mating systems, both the mother and the father are involved in parental care. By contrast, montane voles and meadow voles are asocial, living mostly in isolated burrows, and are sexually promiscuous. In these species the father makes no contribution to rearing the offspring, and the mother does the bare minimum, abandoning the nest about two and a half weeks after birth.

What in the brain makes some voles monogamous and others promiscuous? Let's consider the male voles first. A clue comes from the distribution of a receptor for the hormone vasopressin. The V1a type of vasopressin receptor shows strikingly different distribution patterns in the brains of monogamous voles (like the prairie vole) as opposed to promiscuous voles (like the montane vole). Among other differences, the promiscuous voles have high levels of V1a in one region, called the lateral septum, and low levels in another, called the ventral pallidum, while the monogamous voles showed the opposite expression pattern. Importantly, the molecular structure of the receptors in the two types of vole is nearly identical; it's just the distribution pattern that's different, as determined by the regulatory region of the V1a receptor gene.

There are several reasons to believe that vasopressin signaling is critical in determining the initial partner preference of male rodents. In the male prairie vole, mating with a female stimulates a surge of vasopressin release in the male's brain. Larry Young and his colleagues at Emory University showed that if a drug is used to block the effects of vasopressin on V1a receptors in the ventral pallidum, the male will not form a pair-bond with the female or engage in subsequent parental care. Furthermore, if vasopressin is injected into the brains of male prairie voles, but not montane voles, it causes an increase in so-called affiliative behavior toward nearby females (which in a vole means sniffing the rear end). Most impressively, this group used a molecular genetic trick to express V1a receptors in mice in a pattern similar to that found in monogamous voles. Normal male mice, which mate promiscuously, showed no increase in affiliative behavior toward females after receiving a brain injection of vasopressin. However, the mutant mice with monogamous vole V1a receptor distribution (high levels in the ventral pallidum) showed a significant increase in amorous butt-sniffing. This appeared to be a genuine expression of attachment rather than just greater curiosity about new odors, as vasopressin had no effect on the time mutant mice spent investigating a control odor, a lemon-scented tissue.[27]

While initial establishment of partner preference is crucial for pair-bonding in male prairie voles, it is not sufficient. The male not only has to form an attachment with one female, but also must actively reject other females that follow. In monogamous voles this is expressed not as mere indifference but rather as active aggression by the male toward new females. Some exciting recent work has shown that there are separate brain-signaling pathways mediating initial partner preference and selective aggression toward new females in male monogamous voles. Not surprisingly, the medial forebrain pleasure circuit is involved in both of these aspects of pair-bond formation.

We've often spoken of how dopamine release from VTA neurons onto their targets in the nucleus accumbens is a key step in pleasure and reward. Within the nucleus accumbens there are two different populations of neurons, one of which expresses D2-type dopamine receptors and projects to the ventral pallidum (among other locations) and another that expresses D1 receptors and projects to other regions. Normally, if a pair of prairie voles are allowed to mate as much as they want for twenty-four hours, the males will spend most of their time afterward in close side-by-side contact with their chosen females, indicative of the formation of partner preference.

However, if the pair are placed together for a few hours and during this time are prevented from mating by the experimenters, the males do not form a partner preference. But when a male vole receives an injection of a D2-receptor-activating drug into the nucleus accumbens just before this brief chaperoned date, he will form a strong partner preference even though no copulation has occurred. Conversely, if he is injected with a D2-receptor-blocking drug in the nucleus accumbens, he can mate for as long as twenty-four hours but will fail to form a partner preference.

We have seen that D2 receptor activation is necessary for the first stage of male partner preference. While we don't know the particular cellular details of this mechanism yet, the observation that the D2-receptor-containing neurons in the nucleus accumbens send their axons to the ventral pallidum, where the aforementioned crucial vasopressin receptors reside, suggests that the D2 receptors and the vasopressin receptors interact somehow in the ventral pallidum to trigger partner preference. But what about the other side of the pair-bond formation coin: the aggression toward other females?

When a monogamous male vole is injected with a D1-receptor-activating drug in the nucleus accumbens prior to twenty-four hours of unrestrained mating, he fails to form mate preference. Most important, after a monogamous bachelor male vole has mated with a female and established a pair bond, the level of D1 receptor expression in the nucleus accumbens increases dramatically within a few days, suggesting that this change might underlie selective aggression toward new females.

To test this hypothesis, experimenters used a so-called resident-intruder test to assess male vole behavior. If the intruder was the pair-bonded female, the resident male became romantic and engaged in close affiliative behavior. However, when a female stranger vole was introduced, she was attacked by the male. Remarkably, when a D1-receptor-blocking drug was injected into the nucleus accumbens prior to the introduction of a female stranger, the selective aggression was completely blocked. Instead, the male engaged in close contact with the female stranger, as if he had reverted to his bachelor status.[28]

What about the females? In monogamous female voles, unrestrained interaction with a male triggers a surge of oxytocin in the nucleus accumbens. This long, unchaperoned date would normally be sufficient to induce pair-bond formation in the female, but this can be inhibited by local application of drugs that block oxytocin receptors. The source of oxytocin in the nucleus accumbens appears to be the axons of neurons that reside in the hypothalamus, particularly those regions of the hypothalamus that control hormone release by the adjacent pituitary gland. When genetic tricks were used to increase the levels of oxytocin receptors in the nucleus accumbens of female prairie voles, they formed accelerated partner preference (in the "chaperoned date" protocol).[29]

~~~~~

What does this vole research tell us about our own behavior? Human sexual and social behavior shares some similarities with that of rodents, but has some important differences as well. It shows much greater variability and individuality, for example, and is less closely tied to the olfactory system. At present, it is tempting to speculate that those of us with cheatin' hearts might have differences in brain dopamine, vasopressin, or oxytocin signaling when compared to our more faithful friends who have adopted the prairie vole lifestyle. Indeed there are some suggestive initial findings in support of this idea. For example, analysis of genetic variation in the aforementioned VR1a type of vasopressin receptor showed that male carriers of the "344 allele" of this receptor and their wives reported lower marital satisfaction and a higher incidence of marital crisis in the previous year. Young adults with higher levels of circulating oxytocin reported greater bonding with their parents and a lower incidence of depression. A single study has reported that circulating oxytocin is decreased in autistic subjects as compared to age-matched controls, but this finding has yet to be replicated. There's a great deal of promise for the development of a neurochemical model of long-term human attachment, but that promise has yet to be fully realized.

CHAPTER FIVE

GAMBLING AND OTHER MODERN COMPULSIONS

Whether we are diagnosing Internet addicts, gambling addicts, and porn addicts or examining the motivations of chocaholics and shopaholics, our everyday speech has come to promote the idea that one can become addicted to almost any pleasurable activity. Certainly there's a thread of truth in this assumption—compulsive behaviors *can* impact people's lives to varying degrees. But how similar are such behaviors at a purely biological level? Are addictions to video games, gambling, or shopping really like drug or alcohol addiction in terms of brain function? Or are they just convenient examples of metaphoric language? In her book *Desire: Where Sex Meets Addiction*, Susan Cheever gets to the heart of the issue:

> "Addiction" is the buzzword of the twenty-first century. What we call addiction ranges from the seriousness of methamphetamine addiction . . . to people who say casually they are addicted to Starbucks lattes . . . or sleeping on 600-thread-count sheets. In fact, we especially seem to use the word "addiction" for things to

which we are *not* destructively addicted. . . . These are social habits, and we embrace the word "addiction" to describe them; using it erodes its powers and it identifies us as someone serious, but someone who knows when to take things lightly.[1]

While I agree with Cheever that designer-sheet addiction is overstating the case, both compulsive gambling and video game playing do meet many of the formal behavioral definitions of addiction that have been developed by psychologists, and there are certainly cases where people's lives have been severely affected and even destroyed by such activities. However, behavioral addictions don't necessarily have the same life trajectory as addiction to substances (like drugs, alcohol, or food). In fact, recent community-based studies (as opposed to studies of people in treatment, which are not a representative sample) show that about a third of gambling addicts and video game addicts are able to break their addictions within a given year without seeking outside help, something that rarely happens in drug addiction.

At the biological level there is now reason to believe that a broad definition of addiction—one that encompasses drugs, sex, food, gambling, video games, and some other compulsions—is valid. The developing story is that activation and then alteration of the medial forebrain pleasure circuit is the heart of *all* these addictions. Brain imaging studies have revealed that, like certain drugs or orgasm, both gambling and video game playing engage the medial forebrain pleasure circuit and cause dopamine release in VTA target regions. Recall our earlier discussion of patients who are given dopamine receptor agonist drugs to treat Parkinson's disease. While they have an unusually high incidence of compulsive gambling, their strong urge to do so abates when the drug is withdrawn.

In our zeal to fashion an overall theory of pleasure, reward, and addiction, we must be careful not to overgeneralize. After all, we all eat food and have sex, yet most of us don't become food or sex addicts. In the case of drugs, most people who use alcohol or barbiturates or even cocaine do not develop addictions to these substances. Similarly, most people can gamble or play video games occasionally without this behavior becoming compulsive and ruining their lives.[2] Why is this so? What factors in the biology and/or experience of some individuals will turn pleasure into pathology?

～～～～

In his fascinating memoir of his life as a compulsive gambler, *Born to Lose*,[3] Bill Lee describes a trail of pathology stretching back generations. His grandfather sold his father to another family in China to cover a gambling debt. Lee's father was raised by this surrogate family and then emigrated to the United States, where he also gambled compulsively at mah-jongg and pai-gow poker. Growing up in San Francisco's Chinatown, Lee would accompany his father to his gambling dens as a sort of "good luck charm." In elementary school he was already cutting class to gamble for coins or baseball cards, often losing everything. By high school he was hustling pool and playing poker with varying success and running afoul of loan sharks. Still, he was successful in his studies and went on to get a college degree and a series of well-paying jobs in California's Silicon Valley as a skilled manager and "headhunter" for high-tech firms. He married and had a son, Eric. But as his career advanced he began to gamble more often and for higher stakes. He played the stock market, trading in options. He would work all day in Silicon Valley, drive four hours to the Nevada casinos to play blackjack for a few hours, and then drive back half asleep on icy mountain roads to start work the next morning. This recklessness contributed to the end of his marriage and a subsequent bitter custody battle:

As the custody for Eric became extremely contentious, my urges to gamble became stronger and more frequent. Whereas my preoccupation with gambling used to begin a day or two before my next excursion, it began surfacing sooner. Eventually, the urges started almost as soon as I returned home [from the casino]. All I could think about was getting back to the tables. It wasn't about feeling good or having fun; it was more about not feeling bad.[4]

Within a few years Lee was utterly bankrupt, having gambled away his entire life savings and his home. This destructive cycle continued, destroying a second marriage and squandering another small fortune, leading him to the brink of suicide. Lee joined the twelve-step group Gamblers Anonymous but dropped out several times over the course of many years as the urge to gamble became overwhelming. At one point, having gone ninety days without placing a bet, he went to sleep with a sense of accomplishment at having reached this goal. But then, he recalls, "I woke up drenched in sweat and shaking. My urge to gamble left my entire body feeling like one big mosquito bite, and no amount of willpower would have been able to stop me from scratching myself."[5] As of 2005, he had not placed a bet for four years and was an enthusiastic advocate of Gamblers Anonymous.

Bill Lee's heartwrenching story illustrates not only how gambling addiction can destroy lives, but also some general themes of this disease. As his experience suggests, compulsive gambling runs in families and is much more prevalent in men than in women. Almost certainly, an increased risk for compulsive gambling in people who have a close relative with a gambling addiction reflects both nature and nurture. Several studies have examined gambling addiction in male and female twin pairs, comparing monozygotic (identical) and dizygotic (fraternal) twins. These analyses have suggested that inherited factors account for about 35 to 55 percent of the variation in compulsive gambling among men. In women the story is less clear, and while some studies have reported no significant heritability, these analyses are complicated by the much smaller sample of female gambling addicts.[6]

In chapter 3 we discussed how carriers of the TaqIA A1 allele of the D2 dopamine receptor, who have reduced dopamine signaling in VTA target regions, are more likely to struggle with several different substance addictions: food, drugs, alcohol. Carriers of this genetic variant are also at greater risk for behavioral addictions such as compulsive shopping and gambling, as well as attention-deficit hyperactivity disorder (ADHD). Not surprisingly, genetic analysis has revealed a number of other variants (situated in D4 and D1 dopamine receptors and dopamine transporters) that reduce dopamine signaling that are also associated with gambling addiction and one or more of the other addictive behaviors.[7] Such findings confirm what we already know anecdotally: Anyone who has spent even a little time in a casino has seen that nicotine addiction, alcoholism, and compulsive gambling are often concurrent, reflecting a common underlying disorder of the dopamine-using pleasure circuit. Indeed, the rate of alcoholism among compulsive gamblers is about ten times higher and the rate of tobacco use is about six times higher than in the general age-matched population in the United States.

Bill Lee's story points to several other risk factors for gambling addiction. He was exposed to gambling, as both an observer and a participant, when he was very young and quite poor. And though it may seem relatively trivial, gambling opportunities, both legal and illegal, were also easily available to him in the form of the stock market, card rooms, casinos, and so on. Many studies conducted around the world have come to the same conclusion: When legal gambling becomes more readily accessible, the prevalence of gambling addiction increases. Online gambling, now popular throughout much of the world, is ideally suited to foster gambling addiction, as it is can be indulged in twenty-four hours a day with little social constraint, as no one is likely to be present to urge the gambler to stop.

It is also worth noting that, throughout his cycles of gambling and relapse, Lee managed to thrive in his professional life. Indeed, the risk-taking, hard-driving, and obsessive personality traits often found in compulsive gamblers can be harnessed by some to make them very effective in the workplace. Many gambling addicts are among the most successful, productive, and innovative figures in the business world, a profile that contributes to a self-image of being in control and makes them extremely reluctant to seek help, even in dire circumstances.

When we consider the addiction trajectory of a heroin user, we discover remarkable parallels to Bill Lee's story. Tolerance, withdrawal, craving, and relapse are all present. Lee's clear and resonant description of how his addiction eventually drained all the pleasure from gambling, leaving only a raw desire, could just as easily have been written by a heroin or cocaine addict. The slow transition from liking to wanting that he traces is precisely the same phenomenon experienced by a drug addict and is likely to represent a similar use-dependent rewiring of the pleasure circuit. Furthermore, just like those of drug or food addicts, Lee's most devastating gambling binges were triggered by unusually stressful situations (initially by his divorce and custody battle and again, years later, when he witnessed a mass murder at his workplace). Even after swearing off gambling and joining Gamblers Anonymous, he relapsed several times—an experience typical of both drug and gambling addicts. In fact, one study in Scotland showed that only 8 percent of attendees at Gamblers Anonymous meetings had completely abstained from gambling one year later.[8]

Despite their similarities, it's easy to imagine that compulsive gambling is somehow less destructive than addiction to drugs. However, there are some respects in which it is worse. Most gambling addicts go deeply into debt, and many wind up committing crimes to cover their losses. The life consequences of a disastrous gambling binge can linger for years. Perhaps this is one reason why the attempted suicide rate for gambling addicts is so very high: about 20 percent for Gamblers Anonymous members, rising to as high as 40 percent for a group of men in a residential treatment program for gambling addicts run by the U.S. Veteran's Administration.[9]

~~~~~

A dollar picked up in the road is more satisfaction to you than the ninety-and-nine which you had to work for, and money won at faro or in stocks snuggles into your heart in the same way.

—Mark Twain

So how does someone learn to like gambling? One model holds that early reward is crucial. If you've never gambled before and sit down in the casino to play a few hands of blackjack, you might lose the first five hands, get frustrated, and walk away. You are left with only negative associations (losing money) with gambling and are therefore less likely to try it again. Alternatively, you might win a hand or two early on, thus positively reinforcing the gambling behavior. A subset of people get a small but noticeable pleasure jolt out of this early success, which increases their risk of developing a gambling addiction as they seek more and more stimulation to achieve a "set point of pleasure."

While on the face of it this model seems reasonable, it's likely to be either incorrect or incomplete. Many people who like gambling or even go on to develop gambling addictions did not have an "early win" experience. Similarly, the vast majority of compulsive lottery players, as only one example, will never win the jackpot in a lifetime of betting. An alternative model that has recently emerged from experiments with monkeys and rats suggests that our brains are hardwired to find certain kinds of uncertainty pleasurable (or "rewarding," as it is termed in the cognitive neuroscience literature).[10]

In experiments conducted by Wolfram Schultz and his coworkers at the University of Cambridge, monkeys were trained to watch a computer screen for visual cues while a lick-tube was placed near them to deliver a drop of sweet sugar syrup. At the same time, electrodes were inserted into the monkey's brain to record the activity of individual neurons in the VTA. On the screen there were lights that would turn on and stay on for about two seconds. When a light appeared on the screen—we'll call it green—it indicated that two seconds later a syrup drop would always be delivered. Another light, this one red, indicated that two seconds later no reward would be given.[11]

Let's work through the experiment by following an individual monkey. (This is a bit complicated, so please follow along using Figure 5.1 as a guide.) In the first trial, no light goes on; the delivery of a syrup drop is rapidly followed by a brief burst of dopamine neuron firing in the monkey's brain: In its untrained state it perceives the syrup drop as intrinsically rewarding. (It should be noted here that dopamine neurons in the VTA and other regions

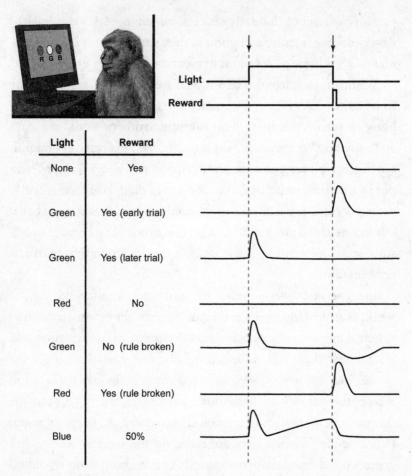

Light	Reward
None	Yes
Green	Yes (early trial)
Green	Yes (later trial)
Red	No
Green	No (rule broken)
Red	Yes (rule broken)
Blue	50%

Figure 5.1 VTA dopamine neurons are activated in anticipation of a reward in the experiments of Schultz and coworkers. See the main text for a complete explanation. Illustration by Joan M. K. Tycko.

are not completely silent at rest. The droplet-evoked burst of spikes is superimposed on a low level of background activity.) The monkey next receives a series of randomly intermixed red-light and green-light trials. In the first few green-light trials, the dopamine neuron fires a burst at the delivery of the syrup droplet, but not at the onset of the green light. But gradually, as the monkey learns that the green light is a reliable predictor of the syrup drop-let, a fascinating change occurs. The dopamine neuron gradually stops responding to the reward itself and instead displays a similar burst of activity at the onset of the green light. The monkey also learns that the red-light trials reliably predict no reward and so red-light trials show no burst activity at any time point.

Let's pause for a moment to consider how amazing this mecha-nism is: The activity of a single VTA dopamine neuron no longer merely indicates simple hardwired pleasure but now represents the learned association between the green light and the sugar droplet reward. While this may seem a trivial point, when pleasure and associative learning are mixed in this way, a minor miracle actu-ally takes place. Now behaviorally compelling stimuli don't have to be intrinsically pleasurable, like sex or food, or artificially plea-surable, like drugs. Any sound, smell, sight, or memory can be-come associated with pleasure and can thereby become pleasurable in its own right.

Back to our monkey story. Next, the experimenters did a clever thing: They broke the rules. For example, in a well-trained mon-key, they flashed the green light but failed to deliver the syrup drop. In this case there was a burst of firing at the green-light onset, but then, two seconds later, when the anticipated syrup droplet failed to arrive, there was a brief decrease in background activity, tempo-rarily driving the neuron to near silence. Alternatively, again using a well-trained monkey, the experimenters flashed the red light, but then violated the learned rule by delivering a syrup drop at red-light offset. This resulted in no burst at red-light onset but a burst immediately following delivery of the unexpected syrup drop. These responses have proven to be extremely useful for guiding learning in the real world. Oftentimes a particular learned associa-tion is no longer valid and has to be overwritten by new experience. In order to do this, the pleasure/reward circuitry of the monkey's brain has to be able to calculate what learning theorists call reward prediction error: the difference between what is expected to happen and what actually happens. Or, in a simple equation:

dopamine neuron response (which encodes reward prediction error) = reward occurrence – reward prediction

When reward occurrence = reward prediction, like in green-light-followed-by-reward trials or red-light-followed-by-nothing trials in well-trained animals, the dopamine neurons fail to burst at light offset. However, when the learned rule is later violated, the dopamine neuron fires a burst at light offset, signaling a reward prediction error. This tells the monkey's pleasure circuit that the old rules don't apply anymore and it may be time to learn a new association.

By now you are probably wondering, with good reason, "What does this have to do with gambling, where the outcome is *always* unpredictable?" In a later experiment, Schultz and his coworkers added yet another visual cue, which we'll call a blue light. The blue light signaled that two seconds after it flashed, a reward would be delivered randomly on 50 percent of the trials. Well-trained animals showed a brief burst of activity when the blue light came on, but then a strange thing happened: In the approximately 1.8-second-long interval between the end of this burst and the offset of the blue light, there was a gradual increase in the level of dopamine neuron firing, such that high rates of firing were achieved by the time the blue light shut off (Figure 5.1, bottom trace). Furthermore, when blue-light trials were presented using extra-large syrup drops, the maximal firing rate achieved during the "waiting interval" was increased.

Essentially, these researchers have created a sort of monkey casino. The period between the onset and the offset of the blue light, when the reward outcome is uncertain, produced a gradually increasing activation of the pleasure circuit in the VTA target regions. This is analogous to the period during which a player is watching the slot machine or the roulette wheel spin or waiting for the turn of a card in blackjack. One reasonable interpretation of these results is that we are hardwired to get a pleasure buzz from risky events. In this model it's not that we need an early reward to like gambling. Rather, the uncertain nature of the payoff is pleasurable in its own right. Evolutionary scenarios have been proposed in which neural systems to drive risk-taking were adaptive, helping an animal beset with indecision to find more reliable predictors of important events. In ancestral humans, this risk-taking may have been more adaptive for male hunters than for female gatherers, potentially underlying the increased risk of present-day males for gambling addiction and other impulse control disorders.

~~~~~

Though monkeys experience a sustained dopamine pleasure buzz in an experiment where the delivery of a syrup drop reward is uncertain, one can take the casino analogy only so far with regard to humans. First, we know that humans have a greatly expanded frontal cortex to guide planning and deciding, a mechanism that might crucially impact the responses to uncertainty. Second, syrup drops are a natural reward, and as we discussed in chapter 3, it's reasonable to conclude that we're hardwired to like sweet foods and drinks. Money, however, is an abstraction, and early ancestral humans certainly didn't use it. Does money really trigger the human pleasure circuit?

Hans Breiter and his coworkers addressed these questions by adapting the monkey protocols of the Schultz lab for use in human brain scanning experiments.[12] Initially each subject received an account containing $50 worth of credit. They were instructed that they were working with real money and that they would be paid the balance of their account in cash at the end of the experiment. In the brain scanner, they watched a video screen that showed one of three wheels, each of which was divided into three pie-shaped segments labeled with a monetary outcome. The "bad" wheel had only negative or neutral outcomes (−$6.00, −$1.50, or $0), an "intermediate" one had mixed results (+$2.50, −$1.50, $0), and a final "good" wheel primarily had rewards (+$10.00, +$2.50, $0). After a particular wheel type was presented on the screen, the subject would push a button that would initiate rotation of an animated pointer. The pointer would spin for about five seconds and then come to rest, seemingly randomly, on one of the three possible outcomes, where it would remain for five more seconds. The design of this experiment makes it possible to measure brain activation during both an anticipation phase (while the pointer is spinning) and an outcome phase (after the pointer has stopped). Of course, the software running the pointer is controlled by the experimenters so that it can deliver all of the possible monetary outcomes in a balanced manner (Figure 5.2).

The main finding was that, as with Schultz's monkeys, VTA target regions (the nucleus accumbens, the orbital gyrus, and the amygdala) were activated during both the anticipation phase and the outcome phase when the outcomes were positive. The anticipation phase responses were graded according to the possible outcome: There was greater activity while the "good" wheel's pointer was spinning than when that of the "intermediate" or "bad" wheel was spinning. And finally, during the outcome phase with the "good" wheel, greatest activation was seen for the largest monetary rewards. Thus even anticipation and experience of an abstract reward, like money, can activate the human pleasure circuit.

This experiment was also designed to test another hypothesis about monetary reward in gambling. Using a related task, Barbara

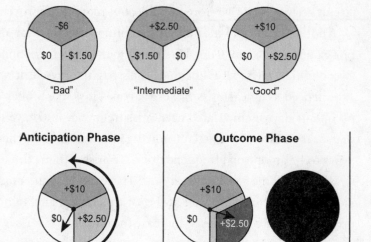

Figure 5.2 The design of an experiment to test the response of human subjects to the anticipation and experience of monetary gains and losses. Adapted from H. C. Breiter, I. Aharon, D. Kahneman, A. Dale, and P. Shizgal, "Functional imaging of neural responses to expectancy and experience of monetary gains and losses," *Neuron* 30 (2001): 619–39, with permission from Elsevier.

Mellers and coworkers demonstrated that people regard a $0 outcome on the "good" wheel as a loss but a $0 outcome on the "bad" wheel as a win.[13] If our minds were completely rational, we would value these outcomes the same way, but we don't. We are influenced by the counterfactual possibility of "what might have been." Was this irrational belief reflected in brain activation? The response strength to the $0 outcome on the "good" wheel was lower than that for the "bad" wheel. However, the responses to the $0 outcome on the "intermediate" wheel did not fall between the levels for the good and the bad $0 responses, as would be predicted. The theory that counterfactual comparison modulates brain pleasure circuit activation is therefore possible, but remains unproven.

Another irrational idea about gambling involves near misses. For example, if a horse one bets on to win comes in second, or if two of three reels on a slot machine's payline match, it will be experienced as a near miss rather than as a loss. A number of experiments have manipulated near-miss frequency and have shown that near misses promote continued gambling. In fact, there appears to be an optimal frequency of near misses to maximally extend slot machine gambling—about 30 percent.[14] Manufacturers of video slot machines are well aware of this effect, and some have programmed their devices to increase the rate of near misses above random levels.[15]

In games of pure chance, like craps or the lottery, gamblers have the same probability of winning whether or not they have a direct involvement in the process (such as buying the lottery ticket or rolling the dice). Nonetheless, many studies have shown that gamblers will bet more and continue gambling longer if they do have a personal role in these fundamentally random events. In some cases, this even affects the style of the particular actions involved in the game. For example, craps players tend to throw the dice with less force when trying to roll low numbers.[16] While both the near-miss effect and the direct-involvement effect are seen in general populations, they are even more prevalent in gambling addicts. Considering these irrational aspects of gambling, Luke Clark and his colleagues at the University of Cambridge hypothesized that there would be significant activation of the pleasure circuit by near misses on a video slot machine and that this activation would be stronger on trials where the gambler had some personal control, as opposed to those presented exclusively by the computer.[17] They placed forty subjects in a brain scanner and presented them with a simplified two-reel video slot machine in which one reel was fixed and the other spun (Figure 5.3). The position of the fixed reel was set by the subject on some trials and by the computer on others. Hits in which the two reels matched yielded a payout of 50 pence. Near misses were those trials in which the matching symbol of the spun reel came to rest either one row above or one row below the payline. Neither near misses nor full misses produced a payout. The computer was programmed to produce near misses on two out of six trials, hits on one out of six trials, and full misses on three out of six trials.

Before each trial the subject was asked, "How do you rate your chances of winning?" After each trial, the subject was asked, "How pleased are you with the result?" and "How much do you

No Win

Total: £0

Figure 5.3 A near miss on a simplified video slot machine in the experiments of Clark et al. (2009). The arrow indicates the payline row. The left reel is fixed on a symbol chosen by either the subject or the computer, and the right reel spins to determine the outcome. From L. Clark, A. J. Lawrence, F. Astley-Jones, and N. Gray, "Gambling near-misses enhance motivation to gamble and recruit win-related brain circuitry," *Neuron* 61 (2009): 481–90, with permission from Elsevier.

want to continue to play the game?" In confirmation of previous findings, personal control of the fixed reel increased both the subject's estimation of his chances and his interest in continuing to play. Also, on winning trials, the pleased-with-result ratings were higher on the personal-control trials as compared with the computer-control trials. When compared to full misses, near misses were experienced as less pleasant but as stimulating the desire to continue to play, but only for those trials where the subject had personal control of the fixed reel.

When the brain scanning data were examined, there were two main findings. First, in all trials, near misses activated much of the same VTA-target pleasure circuit as wins. Both results activated the nucleus accumbens and the anterior insula. However, wins and personal-control trial near misses, but not computer-control trial near misses, also activated another nearby region: the rostral anterior cingulate cortex. These results might help to explain some of the irrational behavior involving gambling: Activation of win-related regions by near-miss outcomes is somehow pleasurable and is more pleasurable when the subject has personal control. This pattern of brain activation could underlie the ability of near misses to promote continued gambling. It's interesting that near-miss outcomes on personal-control trials are simultaneously rated as less pleasant but more compelling to continue. Perhaps this reflects the activation of the pleasure circuit, blended with loss-evoked feelings from other brain regions.

~~~~~

To recap, we know that winning money can activate the human dopamine-using pleasure circuit. We also know that blunted dopamine function in the human pleasure circuit has been found in both drug addicts and food addicts, leading to a suggestion that their addictions result from an attempt to achieve a set point of pleasure that nonaddicts can reach more easily. Could a similar model explain gambling addiction? To test this idea, Christian Büchel and his colleagues at University Hospital Hamburg-Eppendorf in Germany recruited twelve gambling addicts and twelve control subjects to take part in a guessing game with a monetary reward while their brains were scanned.[18] Each subject started with 15 euros and was informed that he would receive the entire balance in cash at the end of the experiment. The simple game involved the presentation of video images of two playing cards, facedown. The subjects were told that one of the cards was red and were asked to guess which by choosing either the right or the left card with a button press. After a two-second delay, the selected card was flipped over. A red card won the subject 1 euro, while a black card lost the subject 1 euro. Of course, the experimenters manipulated the software to control the proportion of wins and losses and their order. The results were engineered so that, at the end of 237 trials, the subject would have a total balance of 23 euros (Figure 5.4).

In both the gambling addicts and the controls, there was significantly greater activation of the nucleus accumbens and the ventrolateral prefrontal cortex (another region that receives VTA dopamine projections) by winning compared to losing. However, when the winning trials were compared in the two groups, the results support the blunted dopamine hypothesis for gambling addiction. Both of these VTA target regions in the gambling addicts were significantly less activated by winning. Interestingly, while this reduction was present on both sides of the gambling addicts' brains, the right side showed a larger reduction than the left. This result is consistent with the findings discussed earlier in which genetic variants that suppress dopamine signaling, particularly in the medial forebrain, were associated with higher rates of gambling addiction.

~~~~~~

While money is not an intrinsic reward in the same way that food, water, and sex are, one could argue that it has come to represent the possibility of intrinsic rewards, and so activation of the pleasure circuit by money is not strictly arbitrary. This begs the question: Can the human pleasure circuit be activated by stimuli that are *entirely* arbitrary? Video games could be a good test case for this question, as they may not provide an intrinsic reward.

Allan Reiss and his coworkers at Stanford University performed brain scanning on subjects playing a simple video game.[19] The subjects were eleven male and eleven female Stanford students, selected to have similar, moderate previous experience with video games and computers generally. The video game involved a screen with a vertical dividing line and leftward-moving balls on the right-hand side, which the player could click to remove (Figure 5.5). When a ball hit the divider, it caused the divider to move slightly leftward, reducing the player's "territory" on the left-hand side of the screen. Conversely, for each second that the area near the divider was kept clear of balls, it would move rightward, gaining territory for the player. The only instruction given was to "click on as many balls as possible." All players soon deduced the point of the game and adopted a click strategy to increase territory.

In all subjects, game play activated a large number of brain regions, including those associated with visual processing, visuospatial attention, motor function, and sensorimotor integration. While these are not surprising results for this task, what was interesting was that key regions of the medial forebrain pleasure circuit were also activated, including the nucleus accumbens, the amygdala, and the orbitofrontal cortex. While both men and women showed activation in these regions during game trials, the effect was significantly stronger in men.

The most provocative aspect of these results is the general finding: Video game play, a completely unnatural behavior divorced from intrinsic reward, activated the pleasure circuit to some degree in all subjects. Perhaps video games tap into some very general pleasure related to goal fulfillment and personal involvement. It's

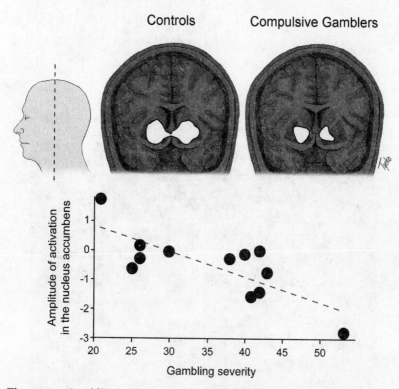

Figure 5.4 Gambling addiction is associated with reduced activation of the medial forebrain pleasure circuit. Top: Brain scan images showing reduced activation of the nucleus accumbens on winning trials in gambling addicts. Bottom: In this graph, each plot point is a different experimental subject, and the scatter plot shows that subjects with the most severe gambling addiction tended to have greater reductions of the pleasure circuit in winning trials. Adapted from J. Reuter, T. Raedler, M. Rose, I. Hand, J. Gläscher, and C. Büchel, "Pathological gambling is linked to reduced activation of the mesolimbic reward system," *Nature Neuroscience* 8 (2005): 147–48, with permission from Macmillan Publishers Ltd., copyright 2005.

also likely that many video games offer a very highly effective reward schedule: Just like cigarettes, the pleasurable moments they provide are brief, but they have rapid onset and are repeated often.

The increased level of activation in men is also interesting, but somewhat harder to interpret. Is there something general about video games that makes them more pleasurable for men? Or is there something about "gaining territory" in a video game that is particularly male-focused? My own suspicion is that the answer

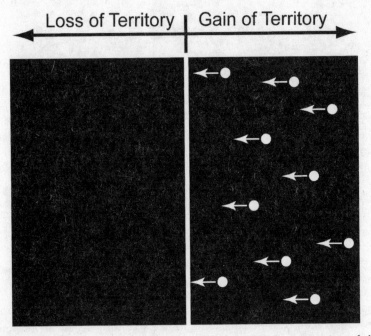

Figure 5.5 The simple video game used to conclude that activation of the midbrain pleasure circuit during video game play is greater in males than females. The player clicked on the balls in the right-hand field in order to move the dividing line to the right and thereby claim more territory. Adapted from F. Hoeft, C. L. Watson, S. R. Kesler, K. E. Bettinger, and A. L. Reiss, "Gender differences in the mesocorticolimbic system during computer game play," *Journal of Psychiatric Research* 42 (2008): 253–58, with permission from Elsevier.

lies in the particular details of the game: If they repeated this study with a combined pattern-recognition and reflex game like Tetris, the gender difference would likely disappear.

An earlier study using a different form of brain scanning (positron emission tomography, or PET) revealed increased dopamine release in subjects playing a tank-driving video game.[20] Furthermore, those subjects who scored highest in the game had the largest dopamine-release signals in the dorsal striatum and nucleus accumbens. While this study is consistent with others demonstrating dopamine pleasure circuit activation in video games, it is complicated by the fact that the subjects were paid (eight UK pounds) for each video game level they completed successfully—thus conflating monetary reward and game play.

If video games can activate the dopamine pleasure circuit, does that mean that one can become addicted to them? The answer seems to be a qualified yes. There is already a burgeoning industry, complete with standardized questionnaires and dubious therapies, that claims to aid in the treatment of video game addiction and Internet addiction.[21] However, media accounts, particularly those originating from East Asia, have overstated both the extent of the problem and its severity. The best indications are that most video game addicts recover without intervention.

~~~~~

This chapter has seen our ideas of the dopamine pleasure circuit extended in some provocative directions. Initially it seemed that the pleasure circuit was either naturally activated by intrinsically adaptive stimuli like food, water, or sex, or artificially engaged by drugs or stimulating electrodes placed deep within the brain. We also discussed how the development of addiction could slowly modify the structure and function of the pleasure circuit and thereby drain the pleasure out of any of these activities, replacing liking with wanting. These observations are all true, but they don't tell the whole story.[22]

We now know from Schultz's monkey experiments that rapid associative learning can transform a pleasure signal into a reward prediction error signal that can guide learning to maximize *future* pleasure. It is likely that this same process is what enables humans to feel pleasure from arbitrary rewards like monetary gain (or even near misses in monetary gain) or winning at a video game. This line of thought leads to some interesting evolutionary and developmental questions. When, exactly, did the ability to feel pleasure from arbitrary rewards develop? And are these rewards really entirely arbitrary, or is there some common theme or quality that runs through them? Can a monkey derive pleasure from playing a video game if there is no intrinsically pleasurable stimulus like a syrup droplet or a jolt of cocaine as the reward? What about a rat? Or a human toddler?

CHAPTER SIX

VIRTUOUS PLEASURES
(AND A LITTLE PAIN)

Jeff Tweedy, leader of the roots-inflected rock bands Wilco and
Uncle Tupelo, struggled mightily with various drug addictions,
most notably to prescription painkillers, alcohol, and cigarettes.
These were coincident with, and in some cases triggered by, chronic
migraines, major depression, and panic attacks that have plagued
him for years. After a successful rehab and several years of drug-
free living, he had this to say about his life:

> I've never felt better. I've never been healthier. . . . I
> run four or five miles, four or five times a week, but
> I broke both my legs running too much last summer.
> I had stress fractures in both my tibias from running
> too much. You know, once you're an addict, you're
> always an addict, so just because I found something
> good to do doesn't mean I'm not going to hurt myself
> doing it.[1]

Yes, as we will discover, exercise *can* activate the pleasure circuit. And so, like nicotine or orgasm or food or gambling, it can become a substrate for addiction as well. This can indeed be a genuine addiction, not merely one as expressed in a common usage like "I'm addicted to sleeping on 600-thread-count sheets." Real exercise addicts display all of the hallmarks of substance addicts: tolerance, craving, withdrawal, and the need to exercise "just to feel normal."[2] Does this make exercise a virtue, a vice, or a little of both?

~~~~~

Sustained physical exercise, whether it be running or swimming or cycling or some other aerobic activity, has well-known health benefits, including improvements in the function of the cardiovascular, pulmonary, and endocrine systems. Voluntary exercise is also associated with long-term improvements in mental function and is the single best thing one can do to slow the cognitive decline that accompanies normal aging. Exercise has a dramatic antidepressive effect. It blunts the brain's response to physical and emotional stress. A regular exercise program produces a large number of changes in the brain, including the new growth and branching of small blood vessels, and increases in the geometric complexity of some neuronal dendrites. It is associated with a host of interrelated biochemical changes as well, including increases in the level of a key protein called BDNF (brain-derived neurotrophic factor). At present we have little understanding of which of these morphological or biochemical changes underlie the beneficial effects of voluntary exercise on brain function, but this is an area of active research.[3]

In addition to the beneficial long-term effects of a sustained exercise program, there are also short-term benefits of exercise that wear off after an hour or two. These include an increased pain threshold, reduction of acute anxiety, and "runner's high."[4] Runner's high (which can occur following any intense aerobic exercise, not just running) is a short-lasting, deeply euphoric state that's well beyond the simple relaxation or peacefulness felt by many following intense exercise. Careful surveys have revealed it to be rather rare: The majority of athletes, whether amateur or professional, never experience it at all, and those who do do so only intermittently. Indeed, many distance runners or swimmers feel merely drained or even nauseated at the end of a long race, not blissful. Since the 1970s runner's high has been attributed in the popular imagination to the exercise-triggered production of endorphins, the brain's own morphine-like molecules. This idea was initiated by a series of studies in which blood was drawn from subjects before and after intense exercise. Analysis revealed an exercise-associated increase in the level of a particular endorphin, called beta-endorphin, in the blood.

There's a major problem, however, in trying to link runner's high with circulating beta-endorphin. Beta-endorphin almost completely fails to cross the cellular barrier that separates the bloodstream from the brain. If beta-endorphin in the bloodstream were indeed responsible for runner's high, then it would have to increase levels of some other chemical messenger that would then cross into the brain to exert its effects. Alternatively, there are different types of endorphins (and related molecules called enkephalins, which together with endorphins are called endogenous opioids) that are synthesized within the brain and therefore could cause euphoria without crossing the blood-brain barrier.

One way to address this question would be to perform a spinal tap on people before and after exercise to see if opioid levels rose in the cerebrospinal fluid that bathes the brain and spinal cord. However, because a spinal tap is painful and carries a small risk of complications, human subjects review boards at most institutions have ruled that it is not ethical to conduct that type of experiment. Dr. Henning Boecker and his coworkers at the University of Bonn in Germany realized that they could investigate runner's high without resorting to spinal taps by measuring brain opioid levels with a brain scanner.[5] They recruited ten amateur distance runners who had previously reported experiencing runner's high. Each subject received a baseline brain scan using a radioactively tagged drug designed to measure the secretion of all forms of endogenous opioid (it bound to all types of the brain's many opioid receptors) and completed a survey of mood. After the subjects had a two-hour-long run followed by a thirty-minute cool-down period, the brain scan and mood survey were repeated. The researchers found that this long run was associated with increased opioid release in the runners' brains, particularly in the prefrontal cortex (a planning and evaluation center) and the anterior cingulate cortex and insula (which serve to interface pain and pleasure with emotions). In addition, those subjects who reported the highest levels of euphoria after running also had the highest levels of opioid release.

This study is an interesting first step, but much more remains to be done in the area. One useful line of work will be to repeat the experiment using more specific opioid receptor drug probes in an attempt to implicate a particular endogenous opioid in runner's high. Then drugs that block those receptors could be given to see if runner's high could be attenuated. It's likely that runner's high is not entirely mediated by the opioid system: Exercise also increases the levels of endocannabinoids, the brain's natural cannabis-like molecules, in the bloodstream. Unlike beta-endorphin, which cannot readily pass the blood-brain barrier, endocannabinoids easily move throughout the body. Thus exercise-induced increases in endocannabinoid levels in blood are presumably mirrored in the brain and could also contribute to the euphoria of runner's high.[6]

Putting some of the pieces together, we know that intense exercise can bring about short-term euphoria, reduction of anxiety, and increases in pain threshold. This is coincident with increases in the levels of brain opioids and, presumably, endocannabinoids, both of which can produce all of these short-term psychoactive effects. We also know that endocannabinoids and opioids can indirectly activate dopamine cells of the VTA and thereby stimulate the medial forebrain pleasure circuit. We know that exercise can be addictive and that other substances and behaviors that are addictive have increased dopamine release in VTA target regions as a common property. In rats, sustained wheel-running can cause dopamine release in the nucleus accumbens and other VTA target regions. Rats also show some signs of exercise addiction. For example, they can be trained to work hard (i.e., perform many lever presses) for access to a running wheel.[7]

All these observations taken together suggest that intense exercise will activate release from VTA dopamine neurons, a process that will underlie at least some portion of runner's high. Unfortunately, to date there is little evidence to support this theory in humans. Gene-Jack Wang and his colleagues at Brookhaven National Laboratory used a brain scanner to image dopamine release in the nucleus accumbens and dorsal striatum of twelve subjects before and after thirty minutes of vigorous treadmill running, followed by a ten-minute cool-down period.[8] They found no differences in D2 dopamine receptor occupancy (their measure of dopamine release) associated with this exercise regimen. No mood scale ratings were performed, so we cannot know if these subjects experienced runner's high. It would be useful to repeat this experiment together with mood scale ratings and more intense exercise, as Boecker and coworkers did for their endogenous opioid measurements.

~~~~~

In the late eighteenth century the British philosopher Jeremy Bentham famously proclaimed, "Nature has placed mankind under the governance of two sovereign masters, *pain* and *pleasure*. . . . They govern us in all we do, in all we say, in all we think: every effort we can make to throw off our subjection, will serve but to demonstrate and confirm it."[9] The accumulating neurobiological evidence indicates that Bentham was only half correct. Pleasure is indeed one compass of our mental function, guiding us toward both virtues and vices, and pain is another. However, we now have reason to believe that they are not two ends of a continuum. The opposite of pleasure isn't pain; rather, just as the opposite of love is not hate but indifference, the opposite of pleasure is not pain but ennui—a lack of interest in sensation and experience. You don't have to be a sadomasochistic sex enthusiast to know that pleasure and pain can be felt simultaneously: Recall Boecker's aching but blissful long-distance runners, or women in childbirth. In the lexicon of cognitive neuroscience, both pleasure and pain indicate *salience*, that is, experience that is potentially important and thereby deserving of attention. Emotion is the currency of salience, and both positive emotions like euphoria and love and negative emotions like fear, anger, and disgust signal events that we must not ignore.

You'll recall our discussion in chapter 4 about how epileptic seizures or brain stimulation with electrodes can produce orgasms that are devoid of pleasure or emotional feeling. While we normally experience orgasm (and everything else, really) as a unified perception, these results revealed that orgasm has in fact dissociable sensory/discriminative and pleasurable/emotional components that are mediated by separate brain regions. The same general theme holds true for pain. There is a sensory/discriminative pathway that runs through the lateral portion of the thalamus, far from the midline, and continues to a region of the cortex involved in touch and muscular sensation (called the primary somatosensory cortex). A parallel pathway involved in the emotional sense of pain runs through the medial thalamus and then contacts two emotion centers, the insula and the anterior cingulate cortex. People who sustain damage limited to the lateral pathway will report an unpleasant emotional reaction to a painful stimulus, but will be unable to describe its specific qualities (dull, stabbing, cold, hot, etc.) or to locate the painful region on their body. Selective damage to the lateral, emotional pathway results in the opposite condition, called pain asymbolia, in which people can report the quality and location of a painful stimulus, but it no longer carries any emotional weight. They have normal reflex-withdrawal when subjected to pain (and a normal reflexive facial grimace), yet the pain just doesn't seem to bother them very much.

In conversation we often use terms like "emotional pain" or "painful social situations." Is this just metaphoric language, or do we actually experience certain powerful emotions the way we do physical pain? The accumulating evidence of recent years indicates that emotional pain activates the medial but not the lateral portion of the physical pain pathway. Experiments that have been devised to inflict even mild social pain (like exclusion from a group task or betrayal by a partner in a gambling game) have demonstrated significant activation of the insula and the anterior cingulate cortex. Emotional pain isn't just a metaphor: In terms of brain activation, it partially overlaps with physical pain.[10]

~~~~~

Both animal and human studies have recently revealed a rather peculiar finding regarding pleasure and pain: Dopamine release from neurons of the VTA, the central biochemical event of the pleasure circuit, is also engaged by painful stimuli. Jon-Kar Zubieta and his colleagues from the University of Michigan performed brain scanning to measure dopamine release in subjects who received a painful stimulus produced by continuously injecting a concentrated salt solution into the jaw muscle.[11] This treatment produced a protracted aching-type pain that lasted for about an hour. The control condition consisted of injecting a normal (isotonic) salt solution that is reported not to be painful. All subjects filled out surveys designed to assess both the emotional and the sensory aspects of pain. The main result was that this long-term painful stimulus was associated with increased dopamine release in both the dorsal striatum and the nucleus accumbens. In the nucleus accumbens, the greatest dopamine release was seen in those subjects who reported the highest emotional pain ratings.

How are we to interpret these findings? Isn't the VTA–nucleus accumbens dopamine pathway supposed to be the core of the pleasure/reward circuit? In trying to make sense of this finding, it is worthwhile to remember that today's brain scanning techniques are very crude. Even a single voxel (a three-dimensional pixel) in a brain scanning image is averaging the response of many thousands to millions of neurons spread over the course of many seconds, so that both the spatial and the temporal resolution are very low. What happens when this sort of experiment is repeated in a rat, where electrodes can be inserted into the brain and recordings made from single dopamine neurons in the VTA? When Mark Ungless and his colleagues at Imperial College London recorded activity from single VTA dopamine neurons in response to a brief (four-second-long) painful foot shock, they found a very interesting result.[12] While all of the dopamine neurons in the VTA were activated by rewards (like a syrup droplet), there were two different patterns of response to the painful stimulus. Dopamine neurons in the dorsal portion of the VTA were transiently inhibited by foot shock: Their firing rate decreased below background levels.[13] However, in the ventral portion of the VTA, dopamine neurons were transiently *activated* by the foot shock. Thus there appear to be two parallel circuits in the VTA. The first is the classic pleasure circuit that we have spent so much time discussing, which is activated by pleasure/rewards and inhibited by pain. The second is a "salience circuit" that is activated by both pleasurable and painful stimuli and is tightly coupled to emotional responses.

Does that mean that pain itself is somehow rewarding, or is it merely salient? The answer isn't entirely clear and awaits further research. There are several factors that complicate any interpretations we might now make. For example, all brief painful stimuli eventually end, and the relief from pain we experience as they do end is itself pleasurable. Chronic, long-lasting pain is a somewhat different story, as it is likely to produce long-term changes in the brain's pleasure circuitry through the action of stress hormones. It is tempting to speculate that the addition of pain to pleasure creates a super-salient response in the medial forebrain and that this somehow contributes to the popularity, in some quarters, of sadomasochistic sex, or even of tasty food loaded with chili peppers.

~~~~~

Another virtuous pleasure that is culturally widespread and often linked to spiritual practice is meditation. But what, exactly, *is* meditation? In her book *The Blissful Brain*, Shanida Nataraja offers these criteria: (1) it must involve a specific technique that is both clearly defined and taught (spacing out in the shower doesn't count), (2) it must involve progressive muscle relaxation, (3) it must involve a reduction in logical processing, and (4) it must be self-induced (thereby excluding the use of drugs or hypnosis).[14] The range of techniques that fall within these boundaries is actually quite large. While all meditative practices involve the conscious regulation of attention and emotion, beyond these basic criteria there is substantial variation. As Richard Davidson, who is both a neurobiologist and an experienced practitioner of meditation, has observed, "Meditation refers to a whole family of practices—it's like using the word 'sports.'"[15]

For example, Yoga Nidra meditation—also known as yogic sleep—is a practice in which the meditator becomes a relaxed, dreamy, neutral observer: She experiences a loss of conscious control of her actions, and her mind withdraws from wishing to act. We can contrast this method with Zen Buddhist meditation, which has the goal of "thinking about not thinking" but prescribes a vigilant mental attitude. This is promoted by a specific seated posture with the eyes open. In Zen meditative practice, mental withdrawal from the sensory world and its attendant dreamy quality are actively discouraged.[16] Yet another meditative practice is Loving Kindness–Compassion meditation from the Buddhist tradition, in which the goals are to counteract self-centered tendencies and ultimately, after long training, to feel a generalized, nonreferential compassion for all beings. The practice does not involve focused attention on particular objects, memories, or images.

In recent years, brain scanners have been employed to identify the regional patterns of activation and deactivation in these meditative states. Not surprisingly, given the wide variety of meditative practices, the results have varied.[17] These studies generally compare the same subject in the meditative state versus some control condition. Herbert Benson and his colleagues at Harvard Medical School used a control condition in which subjects were asked to silently generate a random list of animals as a baseline for measuring changes accompanying a form of Kundalini meditation. In this form of meditation the subjects, who were experienced meditators, monitored their breathing and silently recited *"sat nam"* while inhaling and "*wahe guru*" while exhaling. When compared to the control task, meditation was associated with increased activity in a large number of brain regions, including the dorsolateral prefrontal cortex (a judgment and planning center), anterior cingulate cortex (an emotion center), hippocampus, and striatum, possibly including the nucleus accumbens (the resolution of the images in this particular study makes this latter determination difficult). Some changes were also noted in the brain activation pattern from the earlier to the later stages of the meditation session. By contrast, some but not all studies of Zen meditation have reported *decreases* in the activation of the dorsolateral frontal cortex and the anterior cingulate cortex. Furthermore, within the same meditative tradition, sustained practice of meditation over many years can also influence the pattern of brain activation. Richard Davidson and associates at the University of Wisconsin, Madison, found an inverted U-shaped curve in which expert meditators (average of nineteen thousand hours) had more brain activation than novices, but super-expert meditators (average of forty-four thousand hours) had less. These measurements involved a group of cortical regions that seem to be engaged by attentional processes.[18]

While meditation is certainly relaxing and is sometimes described as blissful, does it in fact activate the medial forebrain pleasure circuit? To date, I am aware of only one study that has sought to address the question directly. Hans Lou and his colleagues at the John F. Kennedy Institute in Denmark performed brain scans on experienced practitioners of Yoga Nidra meditation, using listening to speech with eyes closed as a control condition.[19] They found a significant increase in dopamine release in the nucleus accumbens of their meditators. This result is suggestive, but it awaits both confirmation and extension to other forms of meditation.

~~~~~~

Meditation and prayer lie on a continuum of spiritual practice, but they can share certain relaxed and dissociative qualities. What about the brain activation correlates of more transient, intense phenomena such as ecstatic spiritual or mystical experiences? Mario Beauregard and Vincent Paquette of the University of Montréal have sought to explore this question by performing brain scanning on a group of volunteer Catholic Carmelite nuns.[20] The Carmelites are a contemplative order whose members spend most of their days in silent prayer and meditation, the ultimate goal of which is to achieve *unio mystica*, a state where one feels completely united with God. Because most Carmelites experience *unio mystica* only once or twice in a lifetime of contemplation, they can't simply be hooked up to a brain scanner and commanded to summon the state. To get around this problem, the investigators instructed the nuns to remember and relive their most intense mystical experience. As a control condition they were also asked to remember and relive their most intense state of union with another human (while affiliated with the Carmelite Order).

Comparing brain scans in the mystical remembrance versus the control condition showed activity increases in a group of regions including the anterior cingulate cortex, the orbitofrontal cortex, and portions of the parietal cortex, but not core portions of the medial forebrain pleasure circuit. My own feeling about these results is that they actually have little to do with the neural activation pattern during *unio mystica*. Paquette and Beauregard argue that since trained actors can evoke activation of brain regions associated with particular emotions by recalling past emotional experiences, recalling *unio mystica* is a valid mirror of the primary experience. I remain unconvinced that this construction holds true for intense, euphoric experiences. Does doing your best to recall falling in love really feel like falling in love? Does recalling the experience of orgasm feel like orgasm? Of course not. And neither memory produces the patterns of brain activation that are seen in actual scans of subjects who are falling in love or having an orgasm. Given these conditions, it remains an open question whether ecstatic religious experience activates the medial forebrain pleasure circuit.

But it gets worse. In addition to claiming that the remembrance of mystical union produced the same pattern of brain activation as the experience itself, Beauregard, together with the journalist Denyse O'Leary, go on to claim that this experiment makes a strong case for "the existence of the soul." As they explain, "To the extent that spiritual experiences are experiences in which we contact the reality of our universe, we should expect them to be complex."[21] Yes, nuns recalling *unio mystica* did show activation of a large number of brain regions. But if we examined other brain scanning papers in the biomedical literature, we'd find that many of them revealed that numerous brain regions lit up in response to all kinds of stimuli. If I hooked you up to a brain scanner and had you watch a Three Stooges film, you would also show a complex array of activated brain regions. Conversely, if mystical-union-recalling nuns had shown activation of *few* brain regions, would that count as evidence *against* the existence of the soul? Of course not. The soul may or may not exist, but this study by Beauregard and Paquette adds nothing to our understanding of the issue.

~~~~~

When I was in the first grade, I attended an after-school program at the Jewish Community Center in my hometown of Santa Monica, California. In the lobby was a large banner soliciting donations to the United Jewish Appeal that read "Give 'til It Hurts." I didn't understand it and found the whole thing vaguely disturbing, to the point that, whenever possible, I would navigate around the lobby so as to avoid even looking at the banner. Several months later it was replaced with a similar one—same font, same logo—that read, "Give 'til It Feels Good." *Freakin' adults*, I thought. *Why does everything have to be so confusing?*

All this came back to me when I read a recent paper by William Harbaugh and colleagues at the University of Oregon.[22] The goal of their study was to measure activation of the nucleus accumbens, a neural substrate of pleasure/reward, to test economic models related to taxation and charitable giving. One theory holds that some individuals give to charity out of "pure altruism." They feel satisfaction from providing a public good, like assistance to the needy, and they care only about how much benefit is offered and not the process by which it occurs. This model implies that these individuals should get some pleasure even when such a transfer of wealth is mandatory, as in taxation. A second theory, called "warm glow," holds that people like making their own decision to give. They derive pleasure from the sense of agency, in much the same way that people like to roll their own dice while playing craps and pick their own lottery numbers. In this model, mandatory taxation is not expected to produce a "warm glow." A third theory proposes that some people take pleasure in charitable giving because of its enhancement of their social status. They enjoy being regarded as wealthy or generous by their peers. Of course, these theories are not mutually exclusive. Someone could be motivated by pure altruism and the warm glow of agency *and* the desire for social approval.

Harbaugh and his team designed their experiment to address the first two theories, but not the third. They recruited nineteen young women from the area around Eugene, Oregon, and had them perform various economic transactions in a brain scanner. They were instructed that no one, not even the experimenters, would know their choices. (This was true: Their decisions were written directly to computer disk and machine-coded prior to

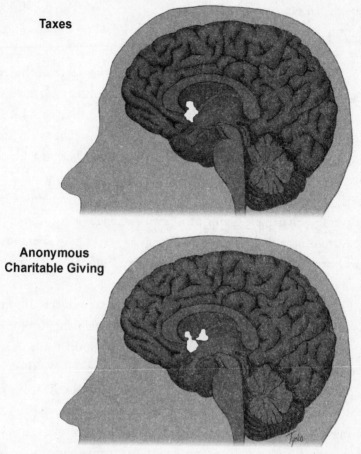

Figure 6.1 The nucleus accumbens, a key part of the medial forebrain pleasure circuit, is activated by both mandatory payouts (taxes; top) and anonymous charitable giving (bottom). These activation patterns show substantial overlap. Adapted from W. T. Harbaugh, U. Mayr, and D. R. Burghart, "Neural responses to taxation and voluntary giving reveal motives for charitable donations," *Science* 316 (2007): 1622–25, with permission from AAAS.

analysis.) Presumably, the design of this experiment removes enhancement of social status as a motivator. Each subject received $100 in an account, which would then be allocated in various amounts to a local food bank. In some of the trials, the subjects had the option to donate; in others they had no choice—they were "taxed." In other trials, they received money with no conditions. The way the study was carried out was as follows: The subjects first were presented an amount of money on a video screen, say $15 or $30. A few seconds later, they learned the status of the trial: This sum was either a gift to them, an involuntary tax on their account, or an offer to donate to charity, which they could either accept or decline by pushing one of two buttons. The brain scanning results showed over the entire population that, just like receiving money, both taxation and charitable giving activated nearly overlapping regions of the nucleus accumbens. However, on average, charitable giving produced a stronger activation of this pleasure center than did taxation (Figure 6.1). These results support both "pure altruism" and "warm glow" models as motivators of charitable giving.

Of course, this doesn't mean that these same subjects are smiling as they write their checks to the IRS, which supports many programs that may be less appealing than a food bank. It also doesn't mean that everyone's brain responds precisely the same way in such conditions. About half of the subjects in the study had more pleasure center activation from receiving money than from giving it, while the other half showed the opposite results. Not surprisingly, those who got more pleasure from giving did indeed choose to give significantly more to charity than the other group. A philosophical question arises from these findings: If giving—even mandatory, anonymous giving—is pleasurable, does that mean that "pure altruism" doesn't really exist? In other words, if we catch a pleasure buzz from our noblest instincts, does that make them less noble?[23]

~~~~~~

While the experiments of Harbaugh and his colleagues masked the subjects' choices in an attempt to eliminate issues of social status and approval, this obviously doesn't reflect conditions in the real world. All of our behavior is embedded in a social context, and this social context powerfully influences our feelings and decisions. We've already mentioned how even mild social rejection can activate the emotional pain centers in the anterior cingulate cortex. Does this mean that positive social interactions can activate pleasure centers as well?

One form of positive social interaction is acceptance—a positive evaluation of the self by others. Norihiro Sadato and his co-workers at the National Institute for Physiological Sciences in Japan have sought to delineate the brain regions activated by one's "good reputation" and compare that to the activation pattern produced by monetary reward. The monetary reward task was like many we have discussed before: Subjects in a brain scanner chose one card of three on a video screen and received various monetary rewards. This produced a pattern of activation similar to that seen in previous studies. The strongest activations were produced by the largest rewards, and these occurred in a number of regions, including the orbitofrontal cortex, the insula, the dorsal striatum, and the nucleus accumbens.

When the same subjects returned for a second day of testing, they took an extensive written personality survey and recorded a short video interview. Then they entered the scanner, where they received social feedback in the form of evaluations of their personality that had supposedly been prepared by a panel of four male and four female observers. To further the deception, they were shown photos of these observers and were told that they would meet them at the end of the experiment. The feedback took the form of a photo of the subject's own face with a single-word descriptor underneath. Some of the descriptors were positive, such as "trustworthy" and "sincere," while others were rather neutral, like "patient." (All of these terms were actually in Japanese, not

English.) Of course, these descriptors were all generated by the experimenters and presented in a randomized order. The main finding was that the most positive social reward descriptors activated portions of the reward circuit—most notably the nucleus accumbens and the dorsal striatum—that substantially overlapped with those activated in the monetary reward task. This finding suggests that there is, quite literally, a common neural currency for social and monetary reward.[24]

~~~~~~

In recent years, many social scientists have come to realize that social comparison can be an important factor in driving economic (and other) decisions of individuals. We evaluate our own economic circumstances and prospects not on some absolute scale, but rather in comparison to those of people around us. We already know that monetary reward can activate certain pleasure centers. If social comparison is indeed encoded in brain function, then it is reasonable to hypothesize that activation of pleasure centers should reflect socially relative rather than absolute levels of payout? To address this issue, Armin Falk and his colleagues at the University of Bonn conducted an experiment in which nineteen pairs of subjects were monitored in side-by-side brain scanners.[25] Each subject simultaneously performed a simple perceptual task: A field of dots was displayed on a video screen for 1.5 seconds, and immediately afterward a number (such as "24") was shown. The subject had to choose, using a rapid button press, whether the dot count was higher or lower than the number. After a short delay, a feedback screen informed the subject about his performance and that of the other subject, along with respective monetary rewards given (e.g., He: 60 euros, You: 120 euros). The subjects received a payout only if they solved the task correctly: If both subjects failed, no payout was given. If only one was correct, he received either 30 or 60 euros, while the other got nothing. However, when both solved the task correctly (which occurred about 66 percent of the time), the computer would assign rewards randomly, from 30 to 120 euros. Sometimes the subjects would get the same reward, sometimes the rewards would be mildly disparate, and other times they would be highly disparate. This experiment showed that social comparison strongly influences activation of reward centers in the brain. The nucleus accumbens was most strongly activated in those trials in which there was a significant difference between one subject's monetary reward and that of the adjacent subject. In other words, despite the biblical injunction of the Tenth Commandment, "Thou shalt not covet thy neighbor's house/wife/slaves/ox/ass/plasma TV/Porsche/etc.," we seem to be hardwired to compare our own experiences and circumstances to those around us.[26]

~~~~~

We humans thrive on information. We love news, gossip, rumors, and, most important, information about our own future. What's more, a number of studies by economists and psychologists have confirmed what we already know from experience: We want that information *now*, not later. Do monkeys also share this desire to know what the future holds? And if so, does this information activate the same VTA dopamine neurons that are stimulated by intrinsically pleasurable stimuli like food and water? In other words, is information about the future pleasurable in and of itself?

These interrelated questions were addressed in a series of experiments performed by Ethan Bromberg-Martin and Okihide Hikosaka at the National Eye Institute in Bethesda, Maryland.[27] They trained two thirsty monkeys to perform a simple decision task: Two targets appeared on the left and right sides of a video screen, and the monkey would choose one by flicking its eyes to a given target. Then, following a delay of a few seconds, the monkey would receive either a large or a small water reward. It didn't matter which target the monkey chose—the rewards were delivered randomly, and at the same overall frequency. The twist to this experiment was that choosing one of the visual targets produced an informative cue during the delay period—a symbol whose shape indicated the size of the upcoming reward—while choosing the other target produced a random cue in the delay period that had no meaning, no predictive value. So in this design, it doesn't matter whether the monkey chooses to receive advance information or a meaningless symbol: It still has the same chance of getting the large water droplet, and it will get it with the same delay.

Just like humans, when given the choice, the monkeys opted to receive information about the future. Within about ten trials, both monkeys were choosing the information-yielding target almost every time. Recordings made from individual dopamine neurons in the VTA and substantia nigra of the subjects revealed that these neurons briefly increased their firing rate when the monkeys saw the symbol that predicted a large amount of water, while the symbol that predicted a small amount of water briefly attenuated their ongoing firing rate. Crucially, after training, these same neurons were excited during probe trials in which the monkey saw only the target that indicated upcoming information and were inhibited when they saw the target that indicated upcoming random, uninformative symbols. The same dopamine neurons that signal the expected amount of pleasure from water also signal the expectation of information, even when that information cannot be put to any use. The monkeys (and presumably humans as well) are getting a pleasure buzz from the information itself.

To my thinking, this experiment is revolutionary. It suggests that something utterly useless and abstract—knowing merely for the sake of knowing—can engage the pleasure/reward circuitry. This is not pleasure obtained from essential things like food or water or sex, which we need to propagate our genes. Nor is it the pleasure of monetary reward, which, while abstract, still represents some real-world benefit, as it can be exchanged for useful things. Nor is it even like the pleasure of charitable giving or the pleasure of receiving positive social feedback, which can also be evolutionarily beneficial for animals living in certain types of social group.

This experiment suggests that ideas are like addictive drugs. As we have seen, certain psychoactive drugs co-opt the pleasure circuit to engage pleasurable feelings normally triggered by food, sex, and so on. In our recent evolutionary lineage (including primates and probably cetaceans), abstract mental constructs have become able to engage the pleasure circuitry as well, a phenomenon that has reached its fullest expression in our own species. The neuroscientist Read Montague, weaving together several strands of thought in cognitive neuroscience from a number of investigators, calls the human ability to take pleasure in abstract ideas a "superpower"[28] and I'm inclined to agree with him. From this perspective, human ideas can even directly oppose our most basic pleasure drives. For example, some people, acting on their religious principles, can forgo sexual activity in service to what they perceive as a more important goal. Likewise, the politically or spiritually motivated hunger-striker is activating her pleasure/reward center by furthering her own ideas, even when this requires acting in precise opposition to one of our most basic and ancient drives.

How does this superpower, in which ideas engage the pleasure circuit, develop on a cellular level? The short answer is that we really don't know. The longer and more speculative answer is that it's just the most recent and elaborate manifestation of the modification of the pleasure circuit by experience (or use-dependent plasticity of neurons, as we like to say in the cellular world). When sensory experience or internal states are represented in our brains by particular patterns of neuronal activity, these patterns of activity can produce changes in neuronal function, particularly electrical function. You'll recall the discussion in chapter 2 of how certain patterns of stimulation can give rise to persistent increases or decreases in the strength of synaptic communication—LTP and LTD, respectively—but LTP and LTD are only one aspect of how experience can change the electrical signaling function of neurons. Furthermore, experience can be written into neuronal memory on different time scales. Some changes can be engaged by a single experience, while others require repetition. Some changes come on rapidly, within seconds, while others require many days. Some of these changes persist only briefly, and others can last a lifetime.

These experience-triggered changes in neuronal function occur throughout the brain, but for the purposes of this discussion, it is critical to emphasize that they take place in the neurons of the medial forebrain pleasure circuit and their immediate connections (the neurons that drive the pleasure circuit and those that are in turn driven by it). What this means is that the simple, hardwired pleasure engaged by ancient stimuli like sex and food can be transformed by experience into much more complex phenomena (Figure 6.2). When Schultz's monkeys learn to associate a green light with an upcoming syrup droplet reward, they rapidly show increased firing of dopamine neurons time-locked to the presentation of this cue. Presumably, there are excitatory axons conveying green-light signals to the dopamine cells of the VTA, and the synapses between these axons and the dopamine cells undergo rapid-onset LTP to create the pleasurable association with the green light.[29]

The same basic model could underlie the association of arbitrary stimuli (like money) or even abstract ideas with pleasure. If one imagines that an abstract idea is represented in the brain by par-

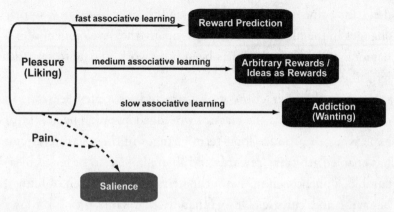

Figure 6.2 The transformation of simple, hardwired pleasure by experience. Pleasure can be transformed through associative learning processes in the medial forebrain circuitry to yield all sorts of phenomena, both beneficial and detrimental. Rapid-onset associative learning in the pleasure circuit can give rise to reward prediction, as with Schultz's monkeys. Repeated association and the attendant slower-onset but longer-lasting changes in circuit function can give rise to arbitrary rewards (like money) and even pleasure driven by ideas, Read Montague's human "superpower." Finally, in some cases, repeated activation of the pleasure circuit by certain drugs or behaviors can give rise to addiction, in which liking is transformed into wanting, and tolerance, withdrawal, and cravings emerge. Illustration by Joan M. K. Tycko.

ticular patterns of neural activity, then those patterns could be conveyed to the pleasure circuit and drive changes within it. It is likely that this form of association would develop more slowly and also be more persistent, like a long-term memory. Long-term memory storage in the brain appears to be associated with microstructural changes in neuronal wiring, and so this form of experience-driven neuronal change is likely to be required for linking abstract ideas to pleasure. Finally, we have seen in chapter 2 how drug addiction can slowly and persistently change the function of the pleasure circuit and thereby transform pleasure and liking into wanting and craving. This is also a long-term process and has been shown to involve changes to neuronal structure, such as increases in dendritic spine density (Figure 2.5). While it is likely that similar changes in the pleasure circuit accompany the development of behavioral addictions, there has been little work to date to test that hypothesis.

In sum, the interaction of pleasure and associative learning in our human brains is the classic two-edged sword: The ability of experience to produce long-term changes in the pleasure circuit has enabled arbitrary rewards and abstract ideas to be felt as pleasurable, a phenomenon that ultimately underlies much of human behavior and culture. Unfortunately, that same process allows pleasure to be transformed into addiction.

CHAPTER SEVEN

THE FUTURE OF PLEASURE

Ray Kurzweil, the noted inventor and futurist, can't wait to get nanobots into his brain. In his view, these devices will be equipped with a variety of sensors and stimulators and will communicate wirelessly with computers outside the body.[1] In addition to providing unprecedented insight into brain function at the cellular level, brain-penetrating nanobots would provide the ultimate virtual reality experience:

> By the late 2020s, nanobots in our brain (that will get there noninvasively, through the capillaries) will create full-immersion virtual-reality environments from within the nervous system. So if you want to go into virtual reality the nanobots shut down the signals coming from your real senses and replace them with the signals that your brain would be receiving if you were actually in the virtual environment. So this will provide full-immersion virtual reality incorporating all of the senses.[2]

Of course, there's no reason why these nanobots must be restricted in their manipulations to the sensory portions of the brain. In Kurzweil's scenario, brain nanobots could just as easily manipulate motor functions, cognitive processes, memories, emotions, and basic drives. Essentially, this idea posits that every neuron in the human brain could have its electrical and chemical activity activated or deactivated with microsecond precision. Every aspect of brain function, from social cognition to regulation of body temperature, could be controlled. Nanobot-driven virtual reality need not be a purely sensory experience. Most germane to our concerns, brain nanobots could manipulate the neurons of the pleasure circuit in extremely precise ways. Do you want a new type of pleasure jolt that's half heroinlike and half gustatory? No problem. Throw in a dash of pain to make it super-salient? Easy as pie.

But nanobot-mediated virtual reality is only the beginning. Kurzweil predicts that by the late 2030s, we will be able to routinely scan an individual's brain with such molecular precision and with such a complete understanding of the rules underlying neuronal function and plasticity that we will be able to "upload" his mental processes into a vastly powerful and capacious future computer. As Kurzweil describes it, "This process would capture a person's entire personality, memory, skills and history."[3] At that point, boundaries between brain, mind, and machine would fall away. Once our individual mental selves are instantiated in machine form, manipulations of mental function, perception, and action just become software modules. Want to improve your mood? Want to preserve all your experiences in memories with perfect fidelity? Want to have the mother of all orgasms? There's an app for that.

As much as I respect Ray Kurzweil and appreciate his willingness to make predictions about and argue for specific future events, I take issue with his timetables for both the introduction of brain-nanobots and the ability to upload the contents and meaning of a brain. The central premise underlying Kurzweil's predictions is that enabling technologies like computer processors, computer memory, microscopes, and DNA sequencing machines have been on an exponential rather than a linear trajectory in terms of their capacity, speed, resolution, and real-world cost, and that it is reasonable to imagine that this exponential trend will continue. Kurzweil also assumes that the human mind resides entirely in the brain (or at least in the nervous system): There is no immortal soul, collective energy, or other nonbiological component that encodes our individual mental selves. At this point in his argument I'm still on board.

However, Kurzweil then argues that our understanding of biology—and of neurobiology in particular—is also on an exponential trajectory, driven by enabling technologies. The unstated but crucial foundation of Kurzweil's scenario requires that at some point in the 2020s, a miracle will occur: If we keep accumulating data about the brain at an exponential rate (its connection patterns, its activity patterns, etc.), then the long-standing mysteries of consciousness, perception, decision, and action will necessarily be revealed. Our understanding of brain function and our ability to measure the relevant parameters of individual brains (aided by technologies like brain nanobots) will consequently increase in an exponential manner to allow for brain-uploading to computers in the year 2039. That's where I get off the bus. I contend that our understanding of biological processes remains on a stubbornly linear trajectory. In my view the central problem here is that Kurzweil is conflating biological data collection with biological insight.

Let's take genetic sequencing as an example. Yes, we have sequenced some human genomes and, yes, the speed and cost of doing so are improving exponentially. The human genome sequence and those of the rat, mouse, fly, and monkey, which have also been completed, are invaluable tools for biologists.[4] That said, while the fundamental insights that have emerged to date from the human genome sequence have been important, they have been far from revelatory. For example, we have learned that gene duplication is more common than we originally thought. And we have learned that humans have fewer genes, but that those genes have more complex modes of regulation and more splice-forms than we had initially predicted. That's all useful information, but it doesn't represent a game-changing, exponential transformation in our understanding of genetics. When the human genome sequence was finished, no one was able to look at it and say, "Aha! now I can understand what makes us uniquely human," or "Aha! now I see how a fertilized egg becomes a newborn during the course of gestation." There *have* been a number of genuine paradigm-shifting insights in genetics in recent years. For example, we now know that chemical modification of DNA through a process called methylation can alter its structure and the way in which it interacts with a set of regulatory/structural proteins called histones, thereby silencing the expression of certain genes. We also know that a set of sequences encoding "micro-RNAs" have powerful roles in determining how other, conventional genes are expressed. Such insights have explained a whole set of puzzles and are a major step forward in our understanding of genetics. But these discoveries, and most of the other key conceptual breakthroughs in this field, have come slowly, the result of stubbornly linear small science, and not of the huge technology-driven data sets that Kurzweil describes.

This linear progress also holds true for the growth in our knowledge of brain function. For example, we now have a map, called the Allen Brain Atlas, that shows the expression pattern of almost every gene in the mouse brain, detailed in a huge series of microscopic images.[5] This resource, which is available to everyone on the Internet, is a great tool for brain researchers, but it has not produced an exponential increase in "Eureka!" moments. The temporal and spatial resolution of our brain scanners is also improving, but these improvements have likewise yielded fundamentally linear insights.

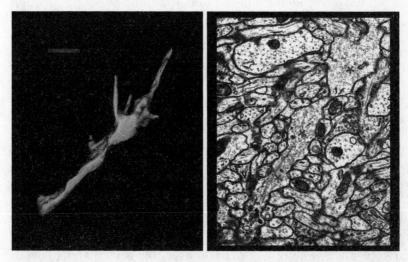

Figure 7.1 The neurons and glial cells of the brain fill almost all of the available space, leaving only minuscule gaps between them. It is easy to look at the image on the left, which shows a computer-based reconstruction of the tip of a growing axon in the brain, and imagine that there is plenty of space around it. However, the complete view of this same growing axon tip is shown in the panel on the right. This image is made with a transmission electron microscope, and it shows how the same growing axon (marked with asterisks) is packed into a dense and complex matrix of tissue containing other neurons and glial cells. The scale bar in the left panel is 0.5 microns long. So you can imagine Kurzweil's brain nanobot, a structure about fourteen times larger in diameter than the scale bar, crashing through this delicate web of living connections. This image is from K. M. Harris, J. C. Fiala, and L. Ostroff, "Structural changes at dendritic spine synapses during long-term potentiation," *Philosophical Transactions of the Royal Society of London, Series B* 358 (2003): 745–48, with permission from the Royal Society.

Don't get me wrong. I do believe that the fundamental and long-standing mysteries of the brain will ultimately be solved. I don't hold with those pessimists who claim that we can never understand our brains by using our brains. I also share Kurzweil's belief that technological advancement will be key to unlocking the enduring mysteries of brain function. But while I see an exponential trajectory in the amount of neurobiological data collected to date, the linear increase in our understanding of neural function means that an idea like brain-penetrating nanobots being usefully deployed by the 2020s seems overly optimistic to me.

Kurzweil's nanobots measure seven microns—about half the diameter of a neuronal cell body—and their job is to maneuver through brain tissue and deploy microsensors and stimulators to evaluate normal brain function. You might imagine the nanobot as functioning like a Volkswagen Beetle: It drives down the road, finds an SUV (a neuron) parked by the side of the road, pulls up next to it, and begins to scan. Here is the first of many problems in Kurzweil's scenario: The brain is composed of neurons and glial cells packed together so tightly that there's almost no room between them (Figure 7.1). What's more, the tiny spaces between these cells are filled not just with salt solution but with structural cables built of proteins and sugars, which have the important function of conveying signals to and from neighboring cells. So let's imagine our nanobot-Volkswagen approaching the brain, where it encounters a parking lot of GMC Yukon SUVs stretching as far as the eye can see. The vehicles are all parked in a grid, with only one half-inch between them, and that half-inch is filled with crucial cables hooked to their mechanical systems. (To be accurate, we should picture the lot as a three-dimensional matrix, a parking lot of SUVs soaring stories into the sky and stretching as far as the eye can see, but you get the idea). Even if our intrepid nanobot were jet-powered and equipped with a powerful cutting laser, how would it move through the brain and not leave a trail of destruction in its wake? It also needs its own power source. And it needs to evade reactive microglia, specialized brain cells that attack and engulf foreign bodies. And all of this has to happen in a way that does not compromise the physiology that the nanobot is trying to measure. These problems are not fundamentally unsolvable, but they are enormous. The 2020s are right around the corner, and there's a lot that must be accomplished in a very short time to keep Kurzweil's nanobot timetable on track.

~~~~~~

Putting aside exotica like brain-uploading and brain-penetrating nanobots for the moment, how can we expect development in our understanding of the neurobiology of pleasure to impact our lives in the next twenty years or so? We biologists are trained to avoid this kind of speculation, as it's simply too easy to get it wrong (Figure 7.2). That said, I'm willing to go out on a limb and offer some ideas about the future of pleasure in both the near term and the more distant future.

One area that is certain to develop in the not too distant future is genetic screening to predict one's risk of developing addictions. You will likely recall that about 50 percent of the variation in the risk for developing various addictions can be accounted for by heredity. In particular, we have seen how genetic variation in the gene encoding the D2 dopamine receptor is correlated with several different forms of addiction: Individuals harboring mutations that decrease the efficacy of D2 receptor function are, on average, more likely to develop addictions to both substances (like alcohol, nicotine, opiates, and food) and behaviors (like compulsive gambling or compulsive sex). There is also evidence that genetic variation in other molecular components of the medial forebrain dopamine-using pleasure circuit is involved. These include other dopamine receptor types, the

Figure 7.2 It's easy to get it wrong when predicting the future of science and technology. The notion of cars powered by nuclear reactors didn't seem entirely unreasonable in 1951. Reprinted with permission from *Motor Trend* (Source Interlink Media, LLC).

dopamine transporter that mediates reuptake into the synaptic terminal of dopamine that has been released into the synaptic cleft (Figure 1.4), and the enzyme COMT (catechol-O-methyltransferase), which breaks down dopamine and some other related neurotransmitters. Genes encoding proteins that are one step removed from the direct action of dopamine can also be significant. For example, activation of dopamine receptors results in the chemical modification of a protein in neurons called DARPP-32, and variation in the DARPP-32 gene is predictive of exploratory behavior, a correlate of addiction. It is likely that determining the genetic sequence of a group of genes involved in medial forebrain dopamine function will provide a much better picture of the risk of developing addictions than would analysis of any single gene.[6]

While the genes involved in dopamine signaling are one key area to look for clues to the heritable nature of addiction, genetic variation in some of the other biochemical systems engaged in various pleasures is also likely to be relevant. These include the endogenous opioids and the endocannabinoids. It will be interesting to see how studies involving them play out in terms of particular addictions and even specific drug addictions. While some genetic variation might be predictive of a generalized addictive predisposition, it is possible that a particular variation in opioid receptors might, for example, correlate with alcohol but not cocaine addiction. (This is pure speculation, intended only to illustrate the general point.) In this vein, it is also worth recalling that there are some biochemical systems that might be engaged by a limited range of pleasurable behaviors. It would not be surprising if variation in the genes encoding oxytocin or its receptors might be relevant for sex addiction, but not other addictions. Similarly, variation in the genes encoding the hormone orexin or the neurotransmitter NPY (or their receptors or effectors), which are critical components of the appetite-regulation circuitry, might be relevant to food addiction, but not to other compulsive behaviors.

Genetic screening is noninvasive—it involves simply swabbing the inside of the cheek to collect some cells—and will become relatively inexpensive once the test is standardized. Screening for addiction risk using brain scanning is much more invasive and costly. Recall a key experiment in chapter 3: Obese subjects showed significantly less activation of the dorsal striatum in response to milkshake sips when compared with the lean subjects, supporting the blunted-pleasure hypothesis of food addiction. It may be that the amplitude of dopamine-mediated responses in regions such as the dorsal striatum and the nucleus accumbens can predict the propensity for a wide variety of addictions. While this approach has the advantage that it could be tailored to check for particular addictions (gambling, food, nicotine), in practice it is a cumbersome and expensive tool that is unlikely to be used widely outside the research laboratory.

One area where there is already some impetus to apply brain scanning of the pleasure circuit is in parole and treatment decisions for convicted male pedophiles. A number of studies have advocated the use of measurements of penile erection to guide these judgments, the rationale being that if, following therapy, a male pedophile still gets an erection in response to photos or videos of nude children, then he is still at risk for committing further sexual abuse.[7] More recently Elke Gizewski and a team at the Swiss Federal Institute of Technology in Zürich have shown that among male homosexual pedophile inpatients (who have yet to receive therapy), photos of nude boys evoked significant activation of the medial forebrain pleasure circuit. This did not occur in a similar population of homosexual men with no signs of pedophilia.[8] Of course, there's a danger of reading too much into both penile erection measures and brain scanning measures in these studies. Almost all of us have been in situations in which we've felt inappropriate sexual arousal but did not act on the feeling. While one wants to do the utmost to protect children from sexual predators, it's not yet clear that brain pleasure circuit activation in response to photos of nude children is a useful predictor of the future behavior of pedophiles.

~~~~~

At the time of this writing, the drugs available to help addicts recover and stay clean are very crude indeed. The most commonly used treatments simply substitute one form of addictive substance for another. Nicotine patches can replace or reduce tobacco smoking. This approach does reduce some smoking-related health problems, but it does not by itself treat the underlying addiction: The subject is still a nicotine addict. Likewise, treatment of heroin addicts with the semi-synthetic opiates methadone and buprenorphine is only a stopgap measure. These drugs act more slowly than heroin and produce less euphoria. They also are delivered orally, thereby eliminating risks associated with injection (like transmission of blood-borne infections). Nonetheless, as with the nicotine patch, the underlying addiction remains untreated, and so the substitute drugs are not a viable long-term solution.

Another strategy has been to design drugs that promote abstinence by creating an aversive response. Disulfiram (marketed as Antabuse), which has been approved as a treatment for alcoholism by the FDA since 1954, works by inhibiting a key enzyme in the multistep breakdown of ethanol, called acetaldehyde dehydrogenase. When this enzyme is blocked, consuming ethanol will produce an increase in levels of acetaldehyde in the bloodstream, which makes the drinker feel very ill.[9] Of course, disulfiram is useful only in carefully monitored settings. Anyone who really wants to drink alcohol will simply discontinue taking the disulfiram pill. And of course, this drug does nothing to blunt the craving for alcohol, but only makes relapse extra painful.

A third strategy has been to develop therapies that prevent a drug of abuse from getting into the brain and thereby exerting its psychoactive effects. One of the most interesting approaches along these lines involves making a vaccine that will engage the patient's own immune system to bind and destroy the drug in the bloodstream, before it enters the brain. Vaccines using nicotine, methamphetamine, or cocaine bound to immune-response-triggering proteins are currently undergoing testing in animal studies.[10] One proposal, fraught with ethical issues, is to offer these vaccines early in life to those with a strong genetic predisposition for addiction.

~~~~~

Only in the last few years have there emerged drugs that actually help blunt the cravings of recovering addicts. One of these is naltrexone (a generic drug, also marketed as Revia), which seems to significantly blunt cravings in abstaining alcoholics. When combined with cognitive behavioral therapy, it produces a significant additional decrease in the rate of relapse. Naltrexone is an antagonist of mu-type opioid receptors and therefore both has a direct action on endogenous opioid signaling and is likely to exert indirect effects on dopamine signaling. The tendency of patients to discontinue taking naltrexone pills can be reduced by use of a newer long-lasting form of naltrexone (sold as Vivitrol), which can be administered as an injection once a month. Naltrexone has also seen some success in reducing relapse in heroin addicts.

Two different drugs have been approved for treating nicotine addiction. Bupropion (sold as Wellbutrin or Zyban) is a dopamine transporter, while varenicline (sold as Chantix or Champix) reduces activation of a particular type of nicotine receptor in the brain.[11] Both of these substances are reported to reduce nicotine cravings in abstaining smokers, and both significantly reduce the rate of smoking relapse (although varenicline seems to be a bit more effective). Unfortunately, both drugs also carry risky side effects, most notably an increase in suicidal thoughts, and so should only be used under the close supervision of a psychiatrist. Mood alteration is a common problem with drugs that target the pleasure circuit: Rimonabant, the appetite-reducing drug that has recently been banned in most of Europe for increased risk of suicide, is a prime example (see chapter 3).

While we are now starting to see the first useful drugs for reducing cravings for nicotine and sedative drugs like alcohol and heroin, almost nothing is available to help those who are trying to abstain from stimulants like cocaine and amphetamines. It's encouraging, therefore, that a large number of new anti-addiction drugs are in various stages of development.[12] While some of these drugs are targeting slightly different aspects of the core biochemical systems of pleasure, like dopamine, opioids, and endocannabinoids, others are branching off into exciting new directions. For example, because we know that stress is a common trigger for relapse in most addictions, including substance addictions like alcohol and behavioral addictions like gambling, one obvious solution is to try to reduce stress through behavioral methods, such as exercise or meditation. Another is to try to interfere with stress hormone action in the brain by using drugs to block receptors for stress hormones such as CRF and neurokinin-1 in the hope that cravings will be reduced. One hypothesis is that stress hormone receptor blockers could prevent the stress-induced LTP of glutamate-using synapses received by VTA dopamine neurons and thereby reduce cravings triggered by pleasure-associated behavioral cues (the sight of the crack pipe, the sound of the slot machine).

The hypothesis that the development of addiction involves slow, persistent changes in the strength and microstructure of glutamate-using synapses within the pleasure circuit suggests that drugs directed against glutamate receptors or the proteins that modulate their function might be useful targets in developing anti-addiction therapies. The problem here is that glutamate is the most widely used neurotransmitter in the brain (and spinal cord), and so drugs that impact glutamate neurotransmission have an unusually large potential for side effects. Fortunately, because there are a wide variety of glutamate receptors, drugs targeting one particular subtype of them have a chance of being useful. This is particularly true of a subset of slow-acting glutamate receptors called metabotropic receptors, which have a more limited distribution in the nervous system and which are engaged only by particular patterns of neural activity.[13] One receptor, called the metabotropic glutamate receptor type 5 (mGluR5), has received particular attention, as it is strongly expressed in key portions of the pleasure circuit, including the neurons of the nucleus accumbens and the dorsal striatum.

When François Conquet and his colleagues at GlaxoSmith-Kline Laboratories in Lausanne, Switzerland, used genetic tricks to create a mouse that lacked mGluR5, they made an astonishing discovery: These mice were utterly indifferent to cocaine. They would not press a lever to self-administer the drug, and they had no special interest in the experimental chamber where cocaine was administered. It's not as if the cocaine had failed to act: Dopamine levels in the pleasure circuit of mGluR5 null mice were still elevated. Rather, it appeared as if the mice simply failed to develop a cocaine addiction.[14] Of course, this result and others like it have helped to spur an enormous interest among drug companies in developing compounds that specifically block or modulate mGluR5. At present, most of these substances are in the preclinical stage: Experiments with rats and mice have suggested that mGluR5 antagonists may hold promise for treating addiction to cocaine, amphetamines, nicotine, and alcohol. As these compounds move into human clinical trials it will become possible to test their effectiveness on behavioral addictions, like gambling, for which there are no viable animal models.[15] It may be that some years hence optimal anti-addiction treatment will involve a combination of drugs to reduce craving (e.g., naltrexone plus an mGluR5 antagonist) together with behavioral therapy.

~~~~~~

When neurophysiologists drift off to sleep at night, they dream of a future in which they can record from and individually stimulate (or inactivate) each of the hundred billion or so neurons in the human brain in completely unconstrained combinations. Unfortunately, this dream is a long way off. For the past sixty years or so, researchers interested in brain systems have generally approached problems in neurophysiology by inserting electrodes in the brains of monkeys and rats and recording the spiking activity of single neurons. This neuron-by-neuron approach has yielded a great deal of information about brain function and continues to do so, but it has many limitations. One of these is that certain types of information in the brain are revealed only when a large number of neurons are recorded simultaneously: They reside in the overall pattern of activity in the population of neurons (measures such as the number and identity of neurons that fire simultaneously, the sequences of firing distributed over a population of neurons, and so on). In the last twenty years, special arrays of closely spaced electrodes have been developed that allow for simultaneous recording from about fifty to two hundred neurons in a particular brain region. In a few cases electrode arrays have even been implanted in two different brain regions simultaneously. These experiments, performed with awake, behaving rats, mice, or monkeys, have revealed some important basic principles of brain function. (For example, neurons in the hippocampus create a spatial map of their environment through path-determined firing patterns; paying attention to a sensory stimulus tends to increases the synchrony of firing in the neurons in the brain that are activated by that stimulus.)[16] It has also been possible to use signals recorded from electrode arrays in the motor cortex to drive robotic limbs, a technique that holds promise for helping people with paralysis from spinal cord injury.[17]

One big problem with either single electrodes or arrays is that it is necessary to drill a hole in the skull of the subject in order to insert them into the brain. Another is that electrodes can trigger potentially damaging inflammatory and microglial cell responses in the brain. As a consequence, they have been used on humans only in very limited circumstances. Today's brain scanners have no such drawbacks, as they can be used repeatedly on humans with no ill effects. But as we have discussed, the information that they provide is very crude. The most common form of brain scanning, called fMRI, does not measure neuronal activity directly. Rather, it measures the regional increase in blood flow that occurs (with about a one-to-three-second delay) when a brain region increases its activity. As of this writing, even the best brain scanners have a spatial resolution of about a millimeter and a temporal resolution of several seconds. Noninvasive brain stimulation or inactivation, using a technique called transcranial magnetic stimulation (TMS), is even less refined: It has a typical spatial resolution of one to two centimeters and cannot effectively act upon deep regions of the brain.

~~~~~~

While we are still very far from achieving the neurophysiologist's ultimate dream of noninvasive, massively parallel single-neuron recording and stimulation in the human brain, a number of current optical technologies for imaging and stimulation point the way toward the future, in general terms. The main challenge in this area is that if you drill a hole in the skull and peer down at the brain's surface, it looks a bit like a boiled egg: It's opaque, not transparent. The brain does not effectively pass light at the wavelengths our eyes can detect. Fortunately, infrared light can penetrate brain tissue somewhat more readily. In a technique known as in vivo multiphoton microscopy, a hole is drilled in the skull (usually of a mouse) and a glass window is cemented in place to cover it. Then, after a week or two in which the mouse is allowed to recover from this surgery, its head is placed under the lens of a microscope that beams extremely bright, brief pulses of light from an infrared laser through the glass window and into the mouse's brain. This light can effectively excite fluorescent molecules at a single focal plane, up to about half a millimeter deep in the brain. These molecules then emit light of a longer, visible wavelength, and this blue, green, or red light is then collected by a sensor attached to the microscope to form a clear image of structures beneath the opaque surface of the living brain (Figure 7.3). Now, I imagine you're thinking, "That's pretty cool, but what does it have to do with the future of pleasure?" Bear with me for just a bit longer and you'll see.

One difficulty with multiphoton microscopy is that relatively few molecules in neurons are naturally fluorescent. In order to see the structure of neurons in the living brain, fluorescent molecules must be artificially introduced.[18] However, one advantage of this technique is that fluorescent probes can be designed to measure different aspects of neuronal structure and function. A molecule that glows constantly and diffuses everywhere in the cell will show the overall outline of the cell and can be used to measure microstructural changes in neurons (as in Figure 7.3). Other molecules can be used to report local calcium ion concentration (which is a proxy for spike activity) or even the electrical potential across the surface of the neuronal membrane. Fluorescent molecules attached to neurotransmitter receptors can be used to measure the number of receptor molecules in a given synapse.

These optical measurement techniques have recently been complemented by another set of tools for controlling the activity of

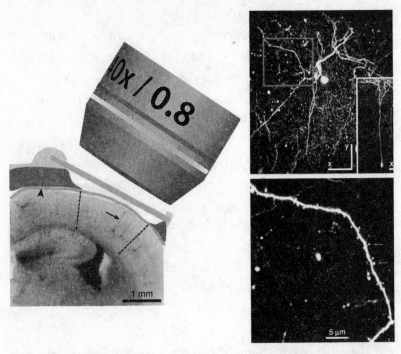

Figure 7.3 In vivo multiphoton microscopy can be used to image structures as far as half a millimeter below the surface of the opaque living brain. The left panel shows the objective lens of the microscope approaching the glass window that has been surgically implanted to cover a small hole in the skull, overlying the tissue of the neocortex. The top right panel shows a top-down view into the neocortex of a mouse that has been genetically engineered to express as fluorescent protein in certain neurons. This image is collapsed in depth so that all of the structures in the field of view have been projected onto the same plane. When the stack of images used to create this projection is rotated in the computer (lower right inset), we can see a reconstructed side view of these neurons. When we zoom in on a section of the projected image (gray box) and expand it, as shown in the lower right panel, we can see living neuronal dendrites studded with spines, where excitatory synapses are received. From A. Holtmaat, T. Bonhoeffer, D. K. Chow, J. Chuckowree, V. De Paola, S. B. Hofer, M. Hübener, T. Keck, G. Knott, W. C. Lee, R. Mostany, T. D. Mrsic-Flogel, E. Nedivi, C. Portera-Cailliau, K. Svoboda, J. T. Trachtenberg, and L. Wilbrecht, "Long-term, high-resolution imaging in the mouse neocortex through a cranial window," *Nature Protocols* 4 (2009): 1128–44, with permission from Macmillan Publishers Ltd., copyright 2009.

neurons with light. Several different laboratories, most notably that of Karl Deisseroth at Stanford University, have developed proteins that can either activate or inhibit neurons very rapidly when they absorb various colors of light.[19] For example, a protein from algae called channelrhodopsin-2 absorbs blue light and then rapidly opens an ion channel that allows the influx of positive ions. When this protein is expressed in neurons (using various genetic tricks), flashes of blue light can then trigger action potentials with millisecond precision. Another microbial protein called halorhodopsin has the opposite effect. When illuminated with yellow light, it will inhibit the ongoing firing of neurons. Neurophysiologists have shown tremendous interest in these tools, and considerable effort is being directed toward optimizing and applying them.

So, here's a summary of the current state of the art: We can use multiphoton microscopy and light-activated molecules to record the electrical activity and the structure of individual neurons, and we can precisely activate and inhibit these neurons in a way that requires that a hole be drilled in the skull to a depth of about half a millimeter but leaves the brain intact. This means that we can access the outer layers of the neocortex, olfactory bulbs, and cerebellum, but not the core pleasure circuitry, emotional centers, hormone control regions, and so on, which are located deep within the brain. If we want to use optical techniques to measure and stimulate neurons in these deeper structures, we must insert an optical fiber to deliver and collect the light—a process that causes significant damage to the brain. In either technique, it is necessary to drill a hole in the skull and to attach bulky auxiliary equipment (microscope, laser, etc.). The fluorescent probes for measurement, or the light-triggered ion channels for optical control of activity, must be delivered through either genetic tricks, viral infection, or in some cases, injection of small molecule probes. There's a bit of a silver lining here, though, in that the genetic tricks or viruses can be constructed to deliver probes to specific subpopulations of neurons.

Different classes of neuron and glial cells are packed very tightly in the brain (Figure 7.1). For example, the VTA dopamine neurons that have been discussed so often are adjacent to glial cells and other non-dopamine VTA neurons. If we are using optical techniques to measure or control the activity of VTA dopamine neurons, then how can we be sure we've got the right cell and not merely its non-dopaminergic neighbor? One strategy would be to try to find a gene that is expressed only in the dopamine neurons (perhaps the dopamine transporter) and use its control elements to force the expression of channelrhodopsin-2 only in those cells. Then, when flashes of blue light are delivered to the VTA, only the dopamine cells will be triggered to fire spikes. This type of strategy, combining genetics and optics, is the current state of the art in terms of the specific control and measurement of identified neurons in the living brain.

~~~~~

Let's play science fiction writer for a moment and imagine a distant future when the neurophysiologist's ultimate dream is actually realized. Perhaps this will involve some form of the technologies that will have evolved from today's multiphoton microscopes and optical controls of neuronal function. You slip on your instrumented baseball cap, which can precisely control the activity of any combination of identified neurons in your brain. What are the possibilities as far as pleasure is concerned? Let's start simply and work our way up. Certainly you could experience a pleasurable sensation similar to that felt by Olds's rats or Heath's patients in response to stimulation of the core pleasure circuit in the medial forebrain (see chapter 1). You could, for example, simulate the pleasures associated with a heroin rush or an orgasm. What's harder to imagine is how the fine control of neural circuitry could give rise to subtler pleasures with new overtones and combinations. For example, we know that heroin activates the dopamine neurons of the VTA (and some other regions). We also know that the dopamine neurons of the VTA project to a number of targets (the nucleus accumbens, the dorsal striatum, the insula, the orbitofrontal cortex, etc.). What would it feel like to stimulate only a subset of these neural projections or a certain group in a certain sequence? An entire parameter space of artificial pleasure could be explored, with a much more subtle and varied group of experiences than can be engaged by drugs like heroin or cocaine or behaviors like sex, eating, or gambling. Imagine that your baseball cap throws up a window in your field of vision, and that this window has a set of sliders you control with your eye movements. You might mix a bit of sexual feeling with some thrill from risk sensation and a bit of food satiety. You could manipulate your emotions in tandem with these pleasures, perhaps adding just a dash of pain into the mix to make it all the more interesting, like eating a good spicy meal. People could share their own recipes for pleasure in much the same way that they now file "trip reports" about psychoactive drugs at www.erowid.org: "Try 2 seconds of high-frequency nucleus accumbens stimulation followed by a sustained oxytocin surge." Because your cap is a recording device as well, you could program it to engage particular pleasure sequences in response to certain real-world (or virtual-world) sensory experiences. And of course, you could choose to network your baseball cap with others to allow for all kinds of interaction.

What about addiction? Wouldn't a large fraction of the population simply become addicted to the cap? Perhaps, but if your cap could control neural circuitry so precisely, it's likely that it could also decouple pleasure and addiction. If addiction is mediated by changes like LTP in the glutamate-using fibers received by dopamine neurons of the VTA or changes in the intrinsic excitability of neurons in the nucleus accumbens, then a stimulation sequence at the end of the session could reset those synapses or those ion channels to counteract the development of addiction. (It might also erase part of the memory of the experience.) Of course, this treatment could be applied to reset the function of the pleasure circuit in conventional (non–baseball cap) addictions as well.

~~~~~~

For the bow cannot stand always bent, nor can human nature or human frailty subsist without some lawful recreation.

—Miguel de Cervantes Saavedra, *Don Quixote de la Mancha*, 1605

Perhaps the hardest thing about imagining the distant future of pleasure is not the technology involved, but rather the social, legal, and financial systems that will surround it. When anyone can precisely control his or her pleasure circuits with an inexpensive head-mounted noninvasive device, how will this ability be used, abused, commercialized, and regulated? If our past experience with psychoactive drugs is any indication, it will be an unholy mess. At various points in our recent past, alcohol, nicotine, heroin, and cocaine have all been legal, banned, taxed, and regulated by the government. Our drug policies reflect a complex web of competing interests. From a capitalist perspective, we have promoted and continue to promote the sale and use of psychoactive drugs because they bring in huge profits. From a medical and social perspective, we seek to ban or limit those psychoactive drugs that cause harm or addiction. But our laws do not always reflect the real risk of various drugs to society: Tobacco is highly addictive, and smoking it kills millions every year, yet it is mostly tolerated. It's almost impossible to kill yourself with cannabis, yet it remains criminalized.[20] I have no faith that the development of future technologies of pleasure will be regulated or commercialized in a more rational way. We will most likely face a politically and commercially driven disaster, as we do with our present drug laws.

If euphoric pleasure is decoupled from addiction, then how will our legal, social, and religious rules for the use of psychoactive drugs change? At present we pay lip service to the notion that addiction is a physiological disease, but most of us continue to harbor the notion that, at some level, addiction represents a failure of will. Most important, this notion informs our medical financial system, in which most insurance carriers will not pay for ongoing addiction treatment, and our legal system, in which addicts are generally punished rather than treated. When the baseball cap of pleasure is in hand, we will need to rethink our moral notions about pleasure. If all kinds of pleasure could be felt without the risk of addiction, then would we still view moderation as a virtue? Would we still consider pleasure to be something that must be earned through labor or sacrifice? These social predictions are perhaps the hardest to make accurately.

In the end, however, thinking about the future of pleasure comes down to the individual. If pleasure is ubiquitous, what will happen to our human "superpower" of being able to associate pleasure with abstract ideas? Will it be washed away in a sea of background noise? If pleasure is everywhere, will uniquely human goals still exist? When pleasure is ubiquitous, what will we desire?

ACKNOWLEDGMENTS

This is my second book and it wouldn't have been written if the first one had simply slipped down the rabbit hole. So, first and foremost I'd like to thank those who read (and bought) my first effort, *The Accidental Mind: How Brain Evolution Has Given Us Love, Memory, Dreams, and God*. Extra thanks to those who took the time to write questions and comments and critique either in published reviews, on the Internet, or in private letters. If not for you, I wouldn't have had the fortitude to try again with the present volume.

Conveying scientific information in a clear and engaging manner is a difficult business and it is my feeling that good illustrations can be invaluable in the struggle. I have been fortunate to partner with Joan Tycko, an intelligent and talented scientific illustrator. Thank you, Joan, for your insight, skill, and great ideas.

A number of people read early drafts of the manuscript and provided excellent feedback to improve it. In particular, heartfelt thanks to David Ginty, Kate Sanford, Elaine Levin, and John Lane. John also provided tremendous support in background research, copyright clearance for images, and good jokes when the times got tough.

Thanks to the members of my laboratory at The Johns Hopkins University School of Medicine for their stimulating ideas and for tolerating my frequent absences from the lab and general spaced-out demeanor in recent months. My colleagues in the Department of Neuroscience have endured way too much lunchtime conversation about my various book-related obsessions, ranging from animal masturbation to psychoactive drugs. Thanks for not having me whacked (always an option in East Baltimore).

This book benefited greatly from the support and critique of publishing industry professionals, most notably my agent, Scott Moyers, who was invaluable in setting the overall trajectory of the book, and my editor, Rick Kot, who pushed me to clarify my thinking and my prose. I'm an established scientist but an amateur writer, and the work that's on these pages owes a lot to the insightful and compassionate efforts of these two professionals.

Even in the best of cases, writing is a lonely and inward-turning process. In this case, it was compounded by an unusually wrenching period in my personal life. Thanks to my children Jacob and Natalie for their unwavering love and support—I couldn't go on without you both. A wonderful circle of friends has been my compass through the storm. In particular, thanks from the depths of my heartmind to Joy McCann, Adam Sapirstein, Kate Sanford, Laura Lipson, and Sascha du Lac.

NOTES

CHAPTER ONE: MASHING THE PLEASURE BUTTON

1. It's fun to go back and read Olds's original descriptions of brain self-stimulation in rats: J. Olds and P. M. Milner, "Positive reinforcement produced by electrical stimulation of septal area and other regions of rat brain," *Journal of Comparative and Physiological Psychology* 47 (1954): 419–27. This review, written a few years after the discovery of the pleasure circuit, is also interesting; J. Olds, "Self-stimulation of the brain: its use to study local effects of hunger, sex, and drugs," *Science* 127 (1958): 315–24.

2. C. E. Moan and R. G. Heath, "Septal stimulation for the initiation of heterosexual behavior in a homosexual male," *Journal of Behavioral Therapy and Experimental Psychiatry* 3 (1972): 23–30. If you want to read about science spinning wildly out of control, you should check out this paper.

3. R. K. Portenoy, J. O. Jarden, J. J. Sidtis, R. B. Lipton, K. M. Foley, and D. A. Rottenberg, "Compulsive thalamic self-stimulation: a case with metabolic, electrophysiologic and behavioral correlates," *Pain* 27 (1986): 277–90.

4. There are several dimensions along which to classify neurotransmitters and their receptors. One of these is "fast versus slow." Both glutamate and GABA, when released from neurons to act on their neighbors, produce very rapid electrical changes: They act within a few thousandths of a second (milliseconds). They do this by binding neurotransmitters that have a pore structure built into them: When the neurotransmitter binds, this causes a subtle change in the shape of the neurotransmitter receptor, opening the pore (called an ion channel) and allowing charged molecules (ions) to flow in or out of the neuron receiving the signal. If the structure of the ion channel allows some ions (like calcium or sodium) to flow, this will increase the

probability that the receiving cell will fire a spike. This is what happens with the fast-acting neurotransmitter glutamate, which we call excitatory because it promotes spike firing. However, when the fast-acting neurotransmitter GABA binds its main receptor, this opens an ion channel that allows chloride ions to flow, producing an electrical effect that suppresses spike firing, and so we call it inhibitory. In addition to fast actions of neurotransmitters like glutamate and GABA, there are also neurotransmitters that have much slower actions. These neurotransmitters bind receptors that don't have ion channels built into them. Rather, they send slow, biochemical signals that affect the electrical behavior of the receiving cell indirectly, often in complex ways that cannot easily be categorized as excitatory or inhibitory. These neurotransmitters, like dopamine and serotonin and norepinephrine, have actions that last from hundreds of milliseconds to tens of seconds.

5. There is a second group of dopamine neurons in a structure adjacent to the VTA called the substantia nigra. These neurons also project to the dorsal striatum, the prefrontal cortex, and the amygdala. They also have a role in pleasure and reward that may be subtly different from the dopamine neurons of the VTA.

6. Drugs that bind to receptors and activate them are called agonists. Drugs that bind to receptors but fail to activate them are called antagonists or blockers. They function to block the activation of these receptors by their natural activators.

7. For a modern analysis of Parkinson's original report, see P. A. Kempster, B. Hurwitz, and A. J. Lees, "A new look at James Parkinson's essay on the shaking palsy," *Neurology* 69 (2007): 482–85.

8. The biomedical literature is full of reports on Parkinson's and pathological gambling. Here's one case history report I adapted: M. Avanzi, E. Uber, and F. Bonfà, "Pathological gambling in two patients on dopamine replacement therapy for Parkinson's disease," *Neurological Science* 25 (2004): 98–101. And here's a nice review of the recent scientific literature on this topic: V. Voon, T. Thomsen, J. M. Miyasaki, M. de Souza, A. Shafro, S. H. Fox, S. Duff-Canning, A. E. Lang, and M. Zurowski, "Factors associated with dopaminergic drug-related pathological gambling in Parkinson's disease," *Archives of Neurology* 64 (2007): 212–16.

9. Of course, the fundamental neurophysiological question, the issue of *qualia*, remains unanswered: Why does dopamine release in the VTA target regions feel pleasurable? A useful discussion of the state of the art may be found in a recent collection of essays: Morten L. Kringelbach and Kent C. Berridge, eds., *Pleasures of the Brain* (Oxford: Oxford University Press, 2010).

CHAPTER TWO: STONED AGAIN

1. Mordecai Cooke, *The Seven Sisters of Sleep* (London: James Blackwood, 1860).

2. A central tenet of the Stoic philosophy of Marcus Aurelius can be found in this line from volume VIII of the *Meditations*: "If thou art pained by any

external thing, it is not this that disturbs thee, but thy own judgment about it. And it is in thy power to wipe out this judgment now." (Translated by George Long.)

3. In the sixteenth century and earlier, sulfuric acid was called vitriol, and so the product formed by mixing sulfuric acid with alcohol to create ether was called sweet vitriol, a term that has an undeniable poetic appeal (and would make a great name for a rock band). Different alcohols mixed with sulfuric acid will create different ethers. Ethanol (a 2-carbon alcohol) will form ethyl ether; methanol (1-carbon alcohol, also called wood alcohol) will form methyl ether, and so on. In the Irish ether-drinking epidemic of the nineteenth century, a mixture of methanol and ethanol, called methylated spirits, was used as the starting material for ether production because, unlike pure ethanol, it was not taxed by the British government. When mixed with sulfuric acid, this produced a mixture of methyl and ethyl ethers in an approximate ratio of 1:6.

4. For contemporaneous accounts of ether drinking in Ireland, see H. N. Draper, "On the use of ether as an intoxicant in the north of Ireland," *Medical Press & Circular* 9 (1870): 117–18; H. N. Draper, "Ether drinking in the north of Ireland," *Medical Press & Circular* 22 (1877): 425–26. Some mostly retrospective reports, written in the waning days of the ether-drinking craze, are E. Hart, "Ether-drinking: its prevalence and results," *British Medical Journal* 2 (1890): 885–90; N. Kerr, "Ether inebriety," *Journal of the American Medical Association* 17 (1891): 791–94. A brief, modern review of the phenomenon is R. A. Strickland, "Ether drinking in Ireland," *Mayo Clinic Proceedings* 71 (1996): 1015.

5. The whole process of handling and drinking ether is extremely dangerous. In 1903, the *British Medical Journal* reported the following terrible accident in the town of Trosno, near Kaliningrad, in present-day Russia (reprinted in *British Medical Journal* 326 [2003] 37): "Ether is drunk by farmers on festive occasions, when it appears to be consumed in pailfuls. A farmer celebrating his son's wedding, in the fullness of his hospitality, got in two pails of ether. During the process of decanting the ether into bottles, a violent explosion took place, by which six children were killed, and one adult was dangerously, and fourteen others more or less severely, injured."

6. "Ayahuasca" is a word in the Quechua language that has been used to describe this preparation. Quechua is not the original language of any of the peoples of the Amazon basin, who have a number of different names for the drink, including "caapi," "yagé," "pildé," and others. The account of Don Emilio Andrade Gomez is adapted from two papers by Dr. Luis Eduardo Luna: "The healing practices of a Peruvian shaman," *Journal of Ethnopharmacology* 11 (1984): 123–33, and "The concept of plants as teachers among four mestizo shamans of Iquitos, northeastern Peru," *Journal of Ethnopharmacology* 11 (1984): 135–56.

7. If you're interested in stoned critters, a good read is Ronald K. Siegel, *Intoxication: The Universal Drive for Mind-Altering Substances* (Rochester, VT: Park Street Press, 2005).

8. The *Amanita* mushroom contains both ibotenic acid and muscimol itself in a ratio of about 10:1. Thus when *Amanita* is eaten, most of the muscimol

that enters the brain does not derive from the mushroom directly, but rather from the enzymatic decarboxylation of ibotenic acid in the body. Muscimol and ibotenic acid have very different biochemical effects on neurons. Muscimol activates receptors for the inhibitory neurotransmitter GABA, while ibotenic acid activates receptors for the excitatory neurotransmitter glutamate. However, it's not clear if this direct action of ibotenic acid on glutamate receptors also contributes to the psychoactive effects produced by the mushroom.

9. You can't imagine how much self-control it has taken not to insert an elaborate joke about "getting pissed" into the narrative here.

10 Siegel, *Intoxication*, p. vii.

11. Griffith Edwards, *Matters of Substance: Drugs—and Why Everyone's a User* (New York: Thomas Dunne/St. Martin's Press, 2005), xix.

12. Acetylcholine is a neurotransmitter that acts on two classes of neurotransmitter receptor. Muscarinic acetylcholine receptors are slow-acting, G-protein-linked receptors, while nicotinic receptors (which are, of course, the ones that are activated by nicotine) are fast-acting because they have an ion channel built into their structure. The synthesis of nicotine by the tobacco plant is part of an ongoing evolutionary story of chemical warfare. Insects also use acetylcholine as a neurotransmitter, and many insects that eat tobacco plants are paralyzed from the ingestion of nicotine. Other insects have evolved behaviors or biological strategies to shield the insect nervous system from nicotine action. (The former include only eating the parts of the plant that have low nicotine levels; the latter include the development of enzymes to rapidly break down nicotine in the insect's body and special cellular sheaths to block nicotine from getting to the neurons.)

13. Ethanol (the common form of alcohol found in alcoholic beverages) has many actions in the brain, and it's not entirely clear which of these are central to alcoholic intoxication. Within the VTA reward circuit, ethanol has direct actions on GABA-A receptors and other ion channels that underlie the production of the spike. There is also reason to believe that ethanol has other pychoactive functions that are entirely independent of dopamine: Dopamine-blocking drugs (or clever genetic engineering tricks in mice to interfere with dopamine receptors) fail to block either ethanol intoxication or ethanol self-administration.

14. The addiction potential of cannabis is an ongoing debate. The best evidence to date is that it does carry some risk of addiction, perhaps similar to that for alcohol. Cannabis addiction research has been hampered by the fact that the active ingredient, THC, as well as synthetic THC-like molecules, are oily and sticky. This makes them hard to inject into the brains of rats or mice, so we have relatively little data from animal models.

15. It is common to speak of coca leaf "chewing" among some people of the Andes, but this is really a misnomer. The traditional mode of use is to mix the coca leaves with wood ashes to render them more alkaline, a process that aids in extraction of the cocaine. Then a wad of treated leaves is packed between the cheek and gum, where, without active chewing, it can deliver a low, steady dose of cocaine for an hour or two.

16. While tobacco smoking has profound health consequences for the smoker,

it involves minimal direct social disruption. For example, unlike heroin addicts, nicotine addicts rarely abandon their children. However, while on a road trip in 1989 I did see a pissed-off mother at a Nebraska truck stop attempt to discipline her rowdy toddlers by blowing cigarette smoke in their faces.

17. The development of the cigarette has played an enormous role in the world-wide spread of nicotine addiction. At the beginning of the nineteenth century, about 60 percent of all tobacco was consumed in the form of snuff (inhaled powdered tobacco), with most of the remainder smoked in pipes or chewed. Cigarettes were hand-rolled creations for most of the nineteenth century, but in 1875 the Allen & Ginter tobacco company offered a prize of $75,000 for a machine to rapidly roll cigarettes. The challenge was taken up by James Bonsack of Roanoke, Virginia, who in 1880 filed a patent for a machine that could roll twelve thousand cigarettes per hour. By 1900 the cigarette had driven snuff to a mere 1 percent of market share and had become the dominant nicotine-dosing device.

18. Thirty years after their initial publication, Lømo and Bliss each wrote brief memoirs about the early days of LTP: T. Lømo, "The discovery of long-term potentiation," *Philosophical Transactions of the Royal Society of London, Series B* 358 (2003): 617–20; T.V.P. Bliss, "A journey from neocortex to hippocampus," *Philosophical Transactions of the Royal Society of London, Series B* 358 (2003): 621–23.

19. S. Schenk, A. Valadez, C. M. Worley, C. McNamara, "Blockade of the acquisition of cocaine self-administration by the NMDA antagonist MK-801 (dizocilpine)," *Behavioral Pharmacology* 4 (1993): 652–59.

20. A good review of the literature on long-term changes in the reward circuitry is J. A. Kauer and R. C. Malenka, "Synaptic plasticity and addiction," *Nature Reviews Neuroscience* 8 (2007): 844–58.

21. If you enjoy tales of famous historical figures engaging in drugs, drink, and other debauchery, I recommend Paul Martin's book *Sex, Drugs and Chocolate* (London: Fourth Estate, 2009).

22. Other complex behavioral traits with a similar degree of heritability (approximately 50 percent), such as childhood shyness or so-called general intelligence, have also not been linked to single genes, and likely reflect polygenic effects.

23. A study conducted by Michael Nader and coworkers at Wake Forest University School of Medicine showed that when cynomolgus monkeys were housed individually, brain scans reveled that they all had similar levels of D2 dopamine receptors in the striatum. However, when twenty of these monkeys were placed together in social housing, species-typical dominance relationships developed. When the social hierarchy was established, the monkeys were rescanned: The dominant monkeys showed a 22 percent increase in D2 receptors, with no significant change in the subordinate monkeys. Later, in a Skinner box, the subordinate monkeys self-administered significantly more cocaine. While development of a social hierarchy isn't a good analog of human talk therapy, the more general point is illustrated: Social experience can drive changes in the function of the pleasure circuit and in addictive behavior. D. Morgan, K. A.

Notes

Grant, H. D. Gage, R. H. Mach, J. R. Kaplan, O. Prioleau, S. H. Nader, N. Buchheimer, R. L. Ehrenkaufer, and M. A. Nader, "Social dominance in monkeys: dopamine D2 receptors and cocaine self-administration," *Nature Neuroscience* 5 (2002): 169–74.

CHAPTER THREE: FEED ME

1. The head/heart duality is a well-known cultural phenomenon. In everyday speech we use "heart" as a shorthand to refer to our emotional state or our faith and "head" to refer to cognition or reason. Should I follow my head or my heart? Both "head" and "heart," while they are literally the names of body parts, are commonly used to stand for nonbodily phenomena, for mental processes. But what body part do we use when we want to refer explicitly to our corporeal self? Why the humble "ass," of course! Consider the seminal gangsta rappers Niggaz with Attitude, who in their classic track "Straight Outta Compton" rhyme: "Niggaz start to mumble / They wanna rumble / Mix 'em and cook 'em in a pot like gumbo / Goin' off on a motherfucker like that / With a gat that's pointed at yo ass." Do the guys in NWA mean to say that a gun is literally pointed downward, at your *tuchus*? Of course not. We understand that in this context "ass" means "corporeal self."
2. The hypothalamus is an interesting part of the brain because it has both neural and endocrine functions. It sends signals in both the conventional neural manner, through spikes that propagate down axons and trigger neurotransmitter release in other brain regions, and in an endocrine fashion, by secreting hormones into the bloodstream. These hormones are distributed throughout the body and so can have wide-ranging effects.
3. The gene that is disrupted in the *db* mouse strain is a bit complicated. Through a process known as mRNA splicing, this one gene actually gives rise to several gene products. They are all members of a family called cytokine receptors. Only one particular splice form, called ObRb, can transduce signals from leptin. While the other products of the *db* gene are distributed widely in the body, the ObRb splice form is strongly enriched in the feeding centers of the hypothalamus, the VTA, and a few other brain regions. For a good review of the leptin and leptin receptor field, see: J. M. Friedman, "Leptin at 14 years of age: an ongoing story," *American Journal of Clinical Nutrition* 89 (2009): 973S–79S.
4. The terms "obese" and "morbidly obese" sound vague and pejorative, but they actually have very particular meanings. Physicians define obesity as a body mass index (BMI) greater than 30 and morbid obesity as a BMI greater than 40. To give you an idea of how this plays out, someone who is five feet nine inches tall would be considered obese at 204 pounds and morbidly obese at 270 pounds.
5. I. S. Farooqi, S. A. Jebb, G. Langmack, E. Lawrence, C. H. Cheetham, A. M. Prentice, I. A. Hughes, M. A. McCamish, and S. O'Rahilly, "Effects of recombinant leptin therapy in a child with congenital leptin deficiency," *New England Journal of Medicine* 341 (1999): 879–84; K. Baicy, E. D. London, J. Monterosso, M. L. Wong, T. Delibasi, A. Sharma, and J. Licinio, "Leptin

replacement alters brain response to food cues in genetically leptin-deficient adults," *Proceedings of the National Academy of Sciences of the USA* 104 (2007): 18276–79.

6. In truth, what we know about the feeding circuit is even more complicated than the stripped-down version I've presented here. For example, NPY isn't the only appetite-stimulating transmitter in the arcuate nucleus (there's another called AGRP), and POMC isn't the only appetite-suppressing transmitter (there's another called CART). Likewise, the lateral hypothalamus doesn't just secrete orexin to stimulate appetite (it also uses another hormone called MCH), and the paraventricular nucleus doesn't only secrete CRH to suppress appetite (it also uses TRH and oxytocin).

7. J. D. Hommel, R. Trinko, R. M. Sears, D. Georgescu, Z. W. Liu, X. B. Gao, J. J. Thurmon, M. Marinelli, and R. J. DiLeone, "Leptin receptor signaling in midbrain dopamine neurons regulates feeding," *Neuron* 51 (2006): 801–10.

8. I. S. Farooqi, E. Bullmore, J. Keogh, J. Gillard, S. O'Rahilly, and P. C. Fletcher, "Leptin regulates striatal regions and human eating behavior," *Science* 317 (2007): 1355.

9. B. M. Geiger, G. G. Behr, L. E. Frank, A. D. Caldera-Siu, M. C. Beinfeld, E. G. Kokkotou, and E. N. Pothos, "Evidence for defective mesolimbic dopamine exocytosis in obesity-prone rats," *FASEB Journal* 22 (2008): 2740–46.

10. A useful review on the parallels between food addiction and drug addiction is N. D. Volkow and R. A. Wise, "How can drug addiction help us understand obesity?" *Nature Neuroscience* 8 (2005): 555–60.

11. E. Stice, S. Spoor, C. Bohon, and D. M. Small, "Relation between obesity and blunted striatal response to food is moderated by *TaqIA* A1 allele," *Science* 322 (2008): 449–52.

12. Eric Stice, interview on National Public Radio, October 16, 2008.

13. While there has been a significant increase in the average weight of citizens in the United States (and many other affluent countries) in recent years, one should be cautious of statistics that tout "an x-fold increase in obesity." It's not that they have the numbers wrong—it's just that the definition of obesity is an arbitrary threshold. If your body mass index (BMI) is 29.9, you're not obese, but if it's 30.1, you are. So small increases in average weight can push a lot of people over the line, artificially amplifying the trend.

14. There are several indications that people whose ancestors were regularly subjected to cycles of famine have a greater propensity to become obese when allowed access to unlimited calories. A discussion of this and related issues may be found in Michael L. Power and Jay Schulkin, *The Evolution of Obesity* (Baltimore: Johns Hopkins University Press, 2009).

15. For an excellent discussion of how corporate test kitchens strive to create craveable foods see Dr. David A. Kessler's book *The End of Overeating* (New York: Rodale, 2009). Dr. Kessler, a former commissioner of the U.S. Food and Drug Administration, has coined the term "conditioned hypereating" to refer to the repeated, compulsive overeating triggered by "hyperpalatable" foods loaded with fat, salt, and sugar.

16. There are at least six different receptors for the neurotransmitter NPY.

However, only two, called NPY1 and NPY5, appear to be involved in stimulation of appetite. Blockers of NPY1 and NPY5 are under investigation as anti-obesity drugs, but it is possible that they might ultimately fail due to side effects on other brain and spinal cord systems in which NPY acts. These include the regulation of blood pressure, pain perception, and insulin secretion.

17. There are different ways in which drugs can produce side effects. One is if the drug is not specific to its molecular target—for example, if rimonabant has effects on neurotransmitter receptors other than CB1. So far, that doesn't seem to be what's happening. The problem with rimonabant is probably more fundamental: It's targeting CB1 specifically, but CB1 is involved in multiple brain systems, including appetite, mood, and nausea. There is one line of hope, however. Rimonabant is a class of CB1-blocker called an inverse agonist. That means it blocks the action of the CB1 receptor in the resting state as well as blocking its action when stimulated by endocannabinoid molecules (or THC in a cannabis smoker). It's possible that other CB1-blocking drugs that are neutral antagonists may have fewer side effects because they do not block the signaling of CB1 in the resting state. If you're interested in the pharmacological details of CB1-acting anti-obesity drugs and the prospects for their further development, you may wish to read the following: D. R. Janero and A. Markiyannis, "Cannabinoid receptor antagonists: pharmacological opportunities, clinical experience, and translational prognosis," *Expert Opinion on Emerging Drugs* 14 (2009): 43–65.

18. While application of leptin alone has had little success in treating obesity, there are some early indications that combination therapy using leptin together with the pancreatic hormone amylin can be useful. In one study, long-term leptin/amylin combination therapy produced a mean weight loss of about 13 percent in obese subjects. This might work because amylin somehow restores leptin responsiveness to the neurons of the hypothalamus.

19. The social rank in small groups of female macaques is not typically enforced by physical violence. Harassment and the threat of physical aggression are sufficient, kind of like middle school. This study is a fascinating read: M. E. Wilson, J. Fisher, A. Fischer, V. Lee, R. B. Harris, and T. J. Bartness, "Quantifying food intake in socially housed monkeys: social status effects on caloric consumption," *Physiology & Behavior* 94 (2008): 586–94.

20. D. Saal, Y. Dong, A. Bonci, and R. C. Malenka, "Drugs of abuse and stress trigger a common synaptic adaptation in dopamine neurons," *Neuron* 37 (2003): 577–82.

21. J. Hahn, F. W. Hopf, and A. Bonci, "Chronic cocaine enhances corticotropin-releasing factor-dependent potentiation of excitatory transmission in ventral tegmental area dopamine neurons," *Journal of Neuroscience* 29 (2009): 6535–44.

22. P. M. Johnson and P. J. Kenny, "Dopamine D2 receptors in addiction-like reward dysfunction and compulsive eating in obese rats," *Nature Neuroscience* 13 (2010): 635–41. For a useful summary and critique of Johnson and Kenny's paper see D. H. Epstein and Y. Shaham, "Cheesecake-eating rats and the question of food addiction," *Nature Neuroscience* 13 (2010): 529–31.

Notes

CHAPTER FOUR: YOUR SEXY BRAIN

1. Orangutans appear to be the runners-up in the long-childhood competition, leaving their mothers at six to eight years of age. However, orangutan dads are solitary creatures and are totally out of the picture in terms of child-rearing.

2. The list of masturbating animals goes on and on: Vampire bats self-stimulate with their feet, walruses with their flippers, kangaroos with their front paws, and savanna baboons with their tails. Let's face it, for both males and females, nature is a huge wank-fest.

3. I refuse to make the obligatory "blowjob" joke here. Science writing is very serious business. And returning to this serious business, a central evolutionary question arises: Why would animals engage in sexual behaviors that don't, at least directly, result in offspring? This question has been addressed in a clearly written recent review article: N. W. Bailey and M. Zuk, "Same-sex sexual behavior and evolution," *Trends in Ecology and Evolution* 24 (2009): 439–46. Bailey and Zuk examine the scientific literature and propose a set of potential explanations of same-sex sexual behavior (which are not mutually exclusive). These include adaptive explanations, such as social glue (in which this behavior reduces tensions, forms alliances, and promotes social cohesion), practice (in which immature individuals learn courtship or mating skills), and kin selection (in which individuals that engage in same-sex sexual behavior provide resources to their siblings or their siblings' off-spring: the doting aunt or uncle effect). Possible nonadaptive explanations include mistaken identity (in which individuals cannot reliably distinguish males from females) and hypersexuality (in which same-sex sexual behavior arises as a by-product when selection acts on a separate but related trait, such as high sexual drive or high sexual responsiveness).

4. The story of the male penguin couple and the female chick they hatched has been made into a lovely children's book, *And Tango Makes Three*, by Justin Richardson and Peter Parnell, illustrated by Henry Cole (New York: Simon & Schuster, 2005). There is also a notable example of long-term female pair-bonding in a wild population. Laysan albatrosses are large seabirds that have established breeding colonies in the Hawaiian Islands. It has recently emerged that about one-third of the breeding pairs in the Oahu colony are two-female situations. These pairs stay together for several years and engage in mutual grooming and copulation. Each typically mates with a male and lays a single egg. But one of these is rolled out of the nest, leaving the other, which hatches to yield the chick the female couple will raise together. To read all about it, see L. C. Young, B. J. Zaun, and E. A. VanderWerf, "Successful same-sex pairing in Laysan albatross," *Biology Letters* 4 (2008): 323–25.

5. C. W. Moeliker, "The first case of homosexual necrophilia in the mallard *Anas platyrhynchos* (Aves:Anatidae)," *DEINSEA* 8 (2001): 243–47.

6. Of course, just because the idea of romantic love exists in a culture, that doesn't mean that it is the main driver of mate selection. There are many cultures where the idea of romantic love is prevalent but is rarely allowed to flourish, as this would subvert existing (male-dominated) religious and social power structures.

7. Psychologists have devised a fifteen-point standardized test called the Passionate Love Scale, or PLS, to quantify these feelings. It contains statements like "I want _____ physically, emotionally, mentally," "Sometimes I can't control my thoughts; they are obsessively on _____," and "I sense my body responding when _____ touches me." The subjects are asked to rank these statements on a nine-point scale from "not true at all" to "definitely true." You can take the test yourself at http://www.pren hall.com/divisions/hss/app/social/chap10_1.html.

8. From the point of the view of the brain scanner, falling in love is not unlike getting high on heroin, cocaine, or amphetamines (thereby validating the premise of a large number of pop songs). Interestingly, lover's-face viewing activated mostly the pleasure circuit on the right side of the brain, while euphoric drugs work on both sides. An unexpected region of activation by love stimuli was the cerebellar deep nuclei, which are involved mostly in the control of motion and motor learning. The work of Brown and coworkers has been reported in A. Aron, H. Fisher, D. J. Mashek, G. Strong, H. Li, and L. L. Brown, "Reward, motivation, and emotion systems associated with early-stage intense romantic love," *Journal of Neurophysiology* 94 (2005): 327–37. This work built upon an earlier study: A. Bartels and S. Zeki, "The neural basis of romantic love," *Neuroreport* 11 (2000): 3829–34. Here I have combined the results of the two studies, which are mostly in agreement.

9. At the time of this writing, these results had not yet been published, but had been communicated informally by Dr. Brown in an online interview at the American Physiological Society. Please see http://www.the-aps.org/press/releases/09/4.htm.

10. This design, in which subjects are prescreened to respond similarly to sexual images, has both advantages and disadvantages. One advantage is that this design controls for average differences in arousal to visual stimuli between men and women. The downside is that the population of women is not entirely representative: It is biased toward those women who are more like men in terms of their response to explicit sexual images. Another issue: On average, are men and women thinking similar thoughts when viewing couple-sex images? Might women be imagining participation to a greater degree? The paper is an interesting read: S. Hamann, R. A. Herman, C. L. Nolan, and K. Wallen, "Men and women differ in amygdala response to visual sexual stimuli," *Nature Neuroscience* 7 (2004): 411–16.

11. Presumably the sports controls were designed to avoid homoerotic content: no images of beefy football players patting each other on the ass. A. Safron, B. Barch, J. M. Bailey, D. R. Gitelman, T. B. Parrish, and P. J. Reber, "Neural correlates of sexual arousal in homosexual and heterosexual men," *Behavioral Neuroscience* 121 (2007): 237–48.

12. The study using the photos of genitals as stimuli was J. Ponseti, H. A. Bosinski, S. Wolff, M. Peller, O. Jansen, H. M. Mehdorn, C. Büchel, and H. R. Siebner, "A functional endophenotype for sexual orientation in humans," *NeuroImage* 33 (2006): 825–33.

13. To my knowledge, this was the first study that combined brain scanning with a measure of genital status. See B. A. Arnow, J. E. Desmond, L. L.

Banner, G. H. Glover, A. Solomon, M. L. Polan, T. F. Lue, and S. W. Atlas, "Brain activation and sexual arousal in healthy, heterosexual males," *Brain* 125 (2002): 1014–23.

14. Here I'm combining the results of three different studies: M. L. Chivers, G. Rieger, E. Latty, and J. M. Bailey, "A sex difference in the specificity of sexual arousal," *Psychological Science* 15 (2004): 736–44; M. L. Chivers and J. M. Bailey, "A sex difference in features that elicit genital response," *Biological Psychology* 70 (2005): 115–20; M. L. Chivers, M. C. Seto, and R. Blanchard, "Gender and sexual orientation differences in sexual response to sexual activities versus gender of actors in sexual films," *Journal of Personality and Social Psychology* 93 (2007): 1108–21. If you're like me, you're probably wondering about the particular content of the videos used in these studies. This is not merely prurient interest. One could imagine that responses to female-female cunnilingus might be different from those to female-female vaginal penetration, for example. Here's the exact description from Chivers et al. (2007):

> The experimental stimuli consisted of 18 film clips that were 90 seconds and that were presented with sound, representing nine stimulus categories: control (landscapes accompanied by relaxing music), nonhuman sexual activity (bonobos or *Pan paniscus* mating), female nonsexual activity (nude exercise), female masturbation, female-female intercourse (cunnilingus and vaginal penetration with a strap-on dildo), male nonsexual activity (nude exercise), male masturbation, male-male intercourse (fellatio and anal intercourse), and female-male copulation (cunnilingus and penile-vaginal intercourse). Participants saw two exemplars of each stimulus category. All of these clips were excerpted from commercially available films.

15. In this study, sexual orientation was assessed by self-report using the Kinsey sexual attraction scale, which uses questions about past sexual behavior to classify respondents along a continuous scale from completely homosexual (score 6) to completely heterosexual (score 0), with gradations in between. The study was G. Rieger, M. L. Chivers, and J. M. Bailey, "Sexual arousal patterns of bisexual men," *Psychological Science* 16 (2005): 579–84.

16. R. J. Levin and W. van Berlo, "Sexual arousal and orgasm in subjects who experience forced or non-consensual sexual stimulation—a review," *Journal of Clinical Forensic Medicine* 11 (2004): 82–88. The conclusion of this review is that neither vaginal lubrication nor orgasm during rape should be taken as an indication of either a woman's feelings of arousal or her consent.

17. This study has important implications for the development of drugs to treat low sex drive in women: The genital vascular response appears to function normally, and so the relevant drug targets are more likely to be in the brain than in the genitals. E. Laan, E. M. van Driel, and R. H. van Lunsen, "Genital responsiveness in healthy women with and without sexual arousal disorder," *Journal of Sexual Medicine* 5 (2008): 1424–35.

18. In one study, women with complete spinal cord lesions achieved orgasm

through vaginal/cervical stimulation while in a brain scanner. The pattern of brain activation indicates that the signals from genital stimulation were conveyed to the brain via the vagus nerve, which exits through the brain stem and is therefore not affected by spinal cord damage. B. R. Komisaruk and B. Whipple, "Functional MRI of the brain during orgasm in women," *Annual Review of Sex Research* 16 (2005): 62–86.

19. As someone who has always enjoyed accounts of alien abduction, I can't tell you how much satisfaction I'm getting by typing the words "rectal probe." For an insightful analysis of the alien abduction phenomenon, I highly recommend Susan A. Clancy, *Abducted: How People Come to Believe They Were Kidnapped by Aliens* (Cambridge: Harvard University Press, 2005).

20. All this makes you wonder what's going on in the mind of the subject. Either she's got to be deep into a faraway fantasy to block out all the unsexy medical stuff, or she's volunteered for this study precisely because she is excited by the rectal probe, the intravenous lines, and the other medical and scientific trappings.

21. Results from several studies are being combined in this account: J. R. Georgiadis, R. Kortekaas, R. Kuipers, A. Nieuwenburg, J. Pruim, A. A. Reinders, and G. Holstege, "Regional cerebral blood flow changes associated with clitorally induced orgasm in healthy women," *European Journal of Neuroscience* 24 (2006): 3305–16; J. R. Georgiadis, A. A. Reinders, A. M. Paans, R. Renken, and R. Kortekaas, "Men versus women on sexual brain function: prominent differences during tactile genital stimulation, but not during orgasm," *Human Brain Mapping* 30 (2009): 3089–101; J. R. Georgiadis, A. A. Reinders, F. H. Van der Graaf, A. M. Paans, and R. Kortekaas, "Brain activation during human male ejaculation revisited," *Neuroreport* 18 (2007): 553–57.

22. Interestingly, all of the various groups of judges were equally clueless. Women were no better than men, gynecologists were no better than psychologists, etc. See E. B. Vance and N. N. Wagner, "Written descriptions of orgasm: a study of sex differences," *Archives of Sexual Behavior* 5 (1976): 87–98.

23. Another class of popular drugs that suppress orgasm are the SSRI antidepressants. Indeed, activation of serotonin receptors, particularly the 5HT-2 subtype, has a powerful orgasm-suppressing and libido-dampening effect (an observation that has led to off-label prescribing of SSRIs for premature ejaculation). Conversely, drugs that have the opposite action, reducing serotonin release or blocking 5HT-2 receptors, stimulate libido and orgasm. How activation of 5-HT2 receptors leads to attenuated orgasm and libido is not entirely clear. One hypothesis has been that serotonin exerts a chronic dampening effect on sexual function through attenuation of dopamine release. However, this is just a hypothesis—it's not even clear that all of the sexual side effects of SSRIs are due to their actions on serotonin levels. It has been suggested that vaginal dryness and dysfunction of penile erection that can result from SSRIs may result from a side effect on a different neurotransmitter acting in the genital tissues, called nitric oxide. One strategy

that some psychiatrists are using to mitigate the sexual side effects of SSRIs is to combine them with the dopamine-boosting drug bupropion (sold as Wellbutrin and Zyban). Another is to combine SSRIs with a 5HT2-receptor-blocking drug called trazodone. This still allows serotonin to act on other types of serotonin receptors to relieve depression. To read more on this topic, see S. H. Kennedy and S. Rizvi, "Sexual dysfunction, depression, and the impact of antidepressants," *Journal of Clinical Psychopharmacology* 29 (2009): 157–64. If you just can't get enough information about orgasm, you may also want to read this book: Barry R. Komisaruk, Carlos Beyer-Flores, and Beverly Whipple, *The Science of Orgasm* (Baltimore: Johns Hopkins University Press, 2006).

24. There's a clearheaded and compassionate examination of sex addiction in Benoit Denizet-Lewis's *America Anonymous: Eight Addicts in Search of a Life* (New York: Simon & Schuster, 2009). This book follows a range of recovering addicts of different backgrounds, ages, and circumstances as they struggle with their addictions to sex, drugs, alcohol, and shoplifting.

25. T. Baumgartner, M. Heinrichs, A. Vonlanthen, U. Fischbacher, and E. Fehr, "Oxytocin shapes the neural circuitry of trust and trust adaptation in humans," *Neuron* 58 (2008): 639–50; M. Kosfeld, M. Heinrichs, P. J. Zak, U. Fischbacher, and E. Fehr, "Oxytocin increases trust in humans," *Nature* 435 (2005): 673–76; G. Domes, M. Heinrichs, A. Michel, C. Berger, and S. C. Herpertz, "Oxytocin improves 'mind-reading' in humans," *Biological Psychiatry* 61 (2007): 731–33.

26. For an interesting review of this field and a discussion of the prospects of certain neurohormone therapies for various personality disorders, see M. Heinrichs, B. von Dawans, and G. Domes, "Oxytocin, vasopressin, and human social behavior," *Frontiers in Neuroendocrinology* 30 (2009): 548–57.

27. M. M. Lim, Z. Wang, D. E. Olazábal, X. Ren, E. F. Terwilliger, and L. J. Young, "Enhanced partner preference in a promiscuous species by manipulating the expression of a single gene," *Nature* 429 (2004): 754–57; M. M. Lim and L. J. Young, "Vasopressin-dependent neural circuits underlying pair bond formation in the monogamous prairie vole," *Neuroscience* 125 (2004): 35–45.

28. B. J. Aragona, Y. Liu, Y. J. Yu, J. T. Curtis, J. M. Detwiler, T. R. Insel, and Z. Wang, "Nucleus accumbens dopamine differentially mediates the formation and maintenance of monogamous pair bonds," *Nature Neuroscience* 9 (2005): 133–39; J. T. Curtis, Y. Liu, B. J. Aragona, and Z. Wang, "Dopamine and monogamy," *Brain Research* 1126 (2006): 76–90.

29. Oxytocin may have a social behavior role in males as well: Both male and female mice in which the oxytocin gene has been deleted do not appear to form social memories of other mice they have encountered previously (they sniff their butts just as assiduously as they would a newly introduced mouse). For a review of oxytocin and pair-bond formation see H. E. Ross and L. J. Young, "Oxytocin and the neural mechanisms regulating social cognition and affiliative behavior," *Frontiers in Neuroendocrinology* 30 (2009): 543–47.

CHAPTER FIVE: GAMBLING AND OTHER MODERN COMPULSIONS

1. Susan Cheever, *Desire: Where Sex Meets Addiction* (New York: Simon & Schuster, 2008), 14–15. Cheever's book is a terrific read, artfully blending stories from her life with information gleaned from interviews with addiction experts ranging from biologists to psychotherapists.

2. The overall rate of gambling addiction has been assessed in many surveys in countries including the United Kingdom, Italy, New Zealand, Sweden, Switzerland, Canada, and the United States. These are complicated by the usual problems involving truthful and complete responses. In addition, some gambling researchers use the term "problem gambling" to refer to a less severe form and "pathological gambling" for a more severe form. Nonetheless, it seems that about 1 to 2 percent of the population in these affluent countries are gambling addicts. A useful meta-analysis of surveys of compulsive gambling is S. Stucki and M. Rihs-Middel, "Prevalence of adult problem and pathological gambling between 2000 and 2005: an update," *Journal of Gambling Studies* 23 (2007): 245–57. A recent telephone survey of fourteen- to twenty-one-year-olds in the United States yielded a compulsive gambling rate of about 2 percent, similar to that seen in adults: J. W. Welte, G. M. Barnes, M. O. Tidwell, and J. H. Hoffman, "The prevalence of problem gambling among U.S. adolescents and young adults: results from a national survey," *Journal of Gambling Studies* 24 (2008): 119–33.

3. Bill Lee, *Born to Lose: Memoirs of a Compulsive Gambler* (Center City, Minn.: Hazelden, 2005).

4. Ibid., 71.

5. Ibid., 145.

6. Two reviews that discuss the genetics of compulsive gambling are N. M. Petry, "Gambling and substance-use disorders: current status and future directions," *American Journal on Addictions* 16 (2007): 1–9; D. S. Lobo and J. L. Kennedy, "The genetics of gambling and behavioral addictions," *CNS Spectrums* 11 (2006): 931–39.

7. Genetic variants that attenuate serotonin signaling have also been implicated in compulsive gambling and ADHD. These include the genes involved in serotonin synthesis (the enzyme tryptophan hydroxylase), serotonin transport, and the enzymatic breakdown of serotonin and other transmitters (monoamine oxidase type A).

8. R. M. Stewart and R. I. Brown, "An outcome study of Gamblers Anonymous," *British Journal of Psychiatry* 152 (1988): 284–88.

9. O. Kausch, "Suicide attempts among veterans seeking treatment for pathological gambling," *Journal of Clinical Psychiatry* 64 (2003): 1031–38.

10. In recent years there has been a slew of general-interest books examining the psychology and neurobiology of human decision-making, what has come to be called neuroeconomics. These include Dan Ariely, *Predictably Irrational: The Hidden Forces That Shape Our Decisions* (New York: Harper-Collins, 2008), and Jonah Lehrer, *How We Decide* (New York: Houghton Mifflin, 2009). While pleasure circuits in the brain are central to decision-making, I'm going to resist the temptation to go much deeper into this material, given the extensive coverage it has already received. If you've enjoyed one or more of these neuroeconomics books and want to delve more deeply

into some of the neurobiology and computational theory underlying this topic, I highly recommend Read Montague, *Why Choose This Book? How We Make Decisions* (New York: Dutton, 2006).

11. In these experiments, recordings were made from both dopamine neurons in the VTA and dopamine neurons in a nearby region called the substantia nigra. The visual signals used by these experimenters were not red, green, and blue lights. They were actually unique geometric patterns. I've just used colors for ease of description. The original, now classic papers are J. R. Hollerman and W. Schultz, "Dopamine neurons report an error in the temporal prediction of reward during learning," *Nature Neuroscience* 1 (1998): 304–9; C. D. Fiorillo, P. N. Tobler, and W. Schultz, "Discrete coding of reward probability and uncertainty by dopamine neurons," *Science* 299 (2003): 1898–902. A more recent and comprehensive review of this topic is W. Schultz, "Multiple dopamine functions at different time courses," *Annual Review of Neuroscience* 30 (2007): 259–88.

12. The paper discussed here is H. C. Breiter, I. Aharon, D. Kahneman, A. Dale, and P. Shizgal, "Functional imaging of neural responses to expectancy and experience of monetary gains and losses," *Neuron* 30 (2001): 619–39. A similar study is B. Knutson, C. M. Adams, G. W. Fong, and D. Hommer, "Anticipation of increasing monetary reward selectively recruits nucleus accumbens," *Journal of Neuroscience* 21 (2001): RC159 (1–5). When comparing the human brain scanner results with microelectrode recordings in monkeys, it should be cautioned that these measures are related but not equivalent. The microelectrode records the spiking activity of individual dopamine neurons in the VTA. The brain scanner is measuring slower changes in blood oxygenation that indirectly indicate the activity of large groups of neurons of many different types in an entire brain region. Microelectrodes have about two thousand times better temporal resolution and about one hundred times better spatial resolution than the brain scanners used in these studies.

13. B. A. Mellers, A. Schwartz, K. Ho, and I. Ritov, "Decision affect theory: emotional reactions to the outcomes of risky option," *Psychological Science* 8 (1997): 423–29.

14. J. I. Kassinove and M. L. Schare, "Effects of the 'near miss' and the 'big win' on persistence at slot machine gambling," *Psychology of Addictive Behaviors* 15 (2001): 155–58.

15. K. A. Harrigan, "Slot machine structural characteristics: creating near misses using high award symbol ratios," *International Journal of Mental Health and Addiction* 6 (2008): 353–68. This study examines a 1989 ruling in which a video slot machine manufacturer was found in violation of Nevada Gaming Commission regulations by programming its terminals to display near misses on the payline in excess of chance levels. Video slot machines also display one row of symbols above and another below the payline, and interestingly, the ruling allowed the continuation of the "virtual reel mapping" technique that employs a different kind of near miss using these rows adjacent to the payline. For example, the payline might display a cherry, a gold bar, and an apple, but the row above the payline would show three plums. This alternate type of near miss also promotes continued gambling.

16. A classic paper relied upon fieldwork conducted at late-night craps games held by off-duty taxi drivers in St. Louis in the 1960s: J. M. Henslin, "Craps and magic," *American Journal of Sociology* 73 (1967): 316–30.

17. L. Clark, A. J. Lawrence, F. Astley-Jones, and N. Gray, "Gambling near-misses enhance motivation to gamble and recruit win-related brain circuitry," *Neuron* 61 (2009): 481–90.

18. The subjects in this study were all men (and almost all of them were tobacco smokers): J. Reuter, T. Raedler, M. Rose, I. Hand, J. Gläscher, and C. Büchel, "Pathological gambling is linked to reduced activation of the mesolimbic reward system," *Nature Neuroscience* 8 (2005): 147–48.

19. F. Hoeft, C. L. Watson, S. R. Kesler, K. E. Bettinger, and A. L. Reiss, "Gender differences in the mesocorticolimbic system during computer gameplay," *Journal of Psychiatric Research* 42 (2008): 253–58.

20. M. J. Koepp, R. N. Gunn, A. D. Lawrence, V. J. Cunningham, A. Dagher, T. Jones, D. J. Brooks, C. J. Bench, and P. M. Grasby, "Evidence for striatal dopamine release during a video game," *Nature* 393 (1998): 266–68.

21. You can see an "Internet Addiction Test," developed by Dr. Kimberly Young, at http://www.netaddiction.com/resources/internet_addiction_test.htm. It has questions like "How often do you lose sleep to late-night log-ins?" There is a large literature in which surveys and interviews are used to assess presumed addiction to video games and various aspects of Internet use. Here is an early example: M. D. Griffiths and N. Hunt, "Dependence on computer games by adolescents," *Psychological Reports* 82 (1998): 475–80. Unfortunately, much of this work is a mess. I must agree with the authors of a recent meta-analysis: S. Byun, C. Ruffini, J. E. Mills, A. C. Douglas, M. Niang, S. Stepchenkova, S. K. Lee, J. Loutfi, J. K. Lee, M. Atallah, and M. Blanton, "Internet addiction: metasynthesis of 1996–2006 quantitative research," *CyberPsychology & Behavior* 12 (2009): 203–7. On page 203 they write, "The analysis showed that previous studies have utilized inconsistent criteria to define Internet addicts, applied recruiting methods that may cause serious sampling bias and examined data using primarily exploratory rather than confirmatory data analysis techniques."

22. Vaughn Bell, writing on the Mind Hacks website, bemoaned an overly simple model of dopamine and pleasure. In so doing, he coined the splendid phrase "four dopamen of the neurocalypse": "My other pet hate is when something pleasurable is described as having the same effect on the brain as one of the four dopamen of the neurocalypse: 'drugs,' 'sex,' 'gambling' and 'chocolate.' Almost any one is used to explain the effect of the others, and if you're really lucky, all four will be invoked to make for an exciting-sounding but often scientifically empty article." http://www.mindhacks.com/blog/2008/02/push_my_brain_button/. This quotation is reproduced here according to the terms of the Creative Commons Attribution License, version 2.0.

CHAPTER SIX: VIRTUOUS PLEASURES (AND A LITTLE PAIN)

1. Jeff Tweedy interview by Jason Crock, http://pitchfork.com/features/interviews/6602-wilco/, posted May 7, 2007.

2. There is a large number of survey-based studies of exercise addiction. One example is V. V. MacLaren and L. A. Best, "Symptoms of exercise dependence and physical activity in students," *Perceptual and Motor Skills* 105 (2007): 1257–64. Not surprisingly, there is also a high rate of exercise addiction among people with body-image-based eating disorders, such as anorexia and bulimia.

3. There is a large body of literature on long-term effects of exercise on the brain and on mental function. A nice recent review that also discusses potential neural synergies between exercise and certain aspects of diet is H. van Praag, "Exercise and the brain: something to chew on," *Trends in Neuroscience* 32 (2009): 283–90.

4. S. Brené, A. Bjørnebekk, E. Åberg, A. A. Mathé, L. Olson, and M. Werme, "Running is rewarding and antidepressive," *Physiology & Behavior* 92 (2007): 136–40; K. F. Koltyn, "Analgesia following exercise," *Sports Medicine* 29 (2000): 85–98.

5. H. Boecker, T. Sprenger, M. E. Spilker, G. Henriksen, M. Koppenhoefer, K. J. Wagner, M. Valet, A. Berthele, and T. R. Tolle, "The runner's high: opioidergic mechanisms in the human brain," *Cerebral Cortex* 18 (2008): 2523–31.

6. A. Dietrich and W. F. McDaniel, "Endocannabinoids and exercise," *British Journal of Sports Medicine* 38 (2004): 536–41.

7. W. M. Wilson and C. A. Marsden, "Extracellular dopamine in the nucleus accumbens of the rat during treadmill running," *Acta Physiologica Scandanavica* 155 (1995): 465–66; I. H. Iversen, "Techniques for establishing schedules with wheel running as reinforcement in rats," *Journal of the Experimental Analysis of Behavior* 60 (1993): 219–38.

8. G. J. Wang, N. D. Volkow, J. S. Fowler, D. Franceschi, J. Logan, N. R. Pappas, C. T. Wong, and N. Netusil, "PET studies of the effects of aerobic exercise on human striatal dopamine release," *Journal of Nuclear Medicine* 41 (2000): 1352–56.

9. Jeremy Bentham, *An Introduction to the Principles of Morals and Legislation* (1789; rev. 1823; reprint, Oxford: Clarendon Press, 1907), 1.

10. I know it looks cheesy to cite your own work, but in this case I'm going to do it anyway. For a useful discussion of emotional versus sensory/discriminative pain circuits in the brain see David J. Linden, *The Accidental Mind: How Brain Evolution Has Given Us Love, Memory, Dreams, and God* (Cambridge, Mass.: Belknap Press of the Harvard University Press, 2007), 100–104.

11. Like a number of other studies we have discussed, this experiment used radioactive raclopride as a tracer, a drug that selectively binds D2-type dopamine receptors. D. J. Scott, M. M. Heitzeg, R. A. Koeppe, C. S. Stohler, and J. K. Zubieta, "Variations in the human pain stress experience mediated by ventral and dorsal basal ganglia dopamine activity," *Journal of Neuroscience* 26 (2006): 10789–95.

12. F. Brischoux, S. Chakraborty, D. I. Brierley, and M. A. Ungless, "Phasic excitation of dopamine neurons in ventral VTA by noxious stimuli," *Proceedings of the National Academy of Sciences of the USA* 106 (2009): 4894–99.

13. You may recall that this is similar to the previously discussed response of dopamine neurons to the absence of an expected reward in Schultz's experiments, trials in a well-trained monkey with a green light followed by no syrup droplet (Figure 5.1).

14. Shanida Nataraja, *The Blissful Brain: Neuroscience and Proof of the Power of Meditation* (London: Gaia, 2008), 18–19. The text in parentheses is my own commentary.

15. This quote is from an interview of Richard Davidson by Bonnie J. Horrigan. It appeared in *Explore* 1 (2005), 380–388.

16. Dögen Kigen, the twelfth-century patriarch of the Soto Zen school in Japan, outlined the practice of *zazen* (seated meditation) as follows: "Think of neither good nor evil and judge not right or wrong. Stop the operation of the mind and consciousness; bring to an end all desires, all concepts and judgments. . . . If a thought arises, take note of it and then dismiss it. When you forget all attachments steadfastly, you will naturally become *zazen* itself." From the text *Fukan Zazengi*, quoted in Hee-Jin Kim, *Eihei Dogen: Mystical Realist* (London: Wisdom Publications, 2004).

17. There is a substantial literature on brain measurements during meditation, most of it performed using the electroencephalograph (EEG), before the advent of brain scanners. A nice review of the literature through 2006 may be found in B. R. Cahn and J. Polich, "Meditation states and traits: EEG, ERP and neuroimaging studies," *Psychological Bulletin* 132 (2006): 180–211. Two exemplars from the meditation and brain scanning literature are S. W. Lazar, G. Bush, R. L. Gollub, G. L. Frichione, G. Khalsa, and H. Benson, "Functional brain mapping of the relaxation response and meditation," *NeuroReport* 11 (2000): 1581–85; G. Pagnoni, M. Cekic, and Y. Guo, "'Thinking about not-thinking': neural correlates of conceptual processing during Zen meditation," *PLoS ONE* 3 (2008): e3083.

18. J. A. Brefczynski-Lewis, A. Lutz, H. S. Schaefer, D. B. Levinson, and R. J. Davidson, "Neural correlates of attentional expertise in long-term meditation practitioners," *Proceedings of the National Academy of Sciences of the USA* 104 (2007): 11483–88.

19. T. W. Kjaer, C. Bertelsen, P. Piccini, D. Brooks, J. Alving, and H. C. Lou, "Increased dopamine tone during meditation-induced change of consciousness," *Cognitive Brain Research* 13 (2002): 255–59.

20. Mario Beauregard and Denyse O'Leary, *The Spiritual Brain: A Neuroscientist's Case for the Existence of the Soul* (New York: HarperCollins, 2007). If you would like to read the key paper that underlies this book, it's M. Beauregard and V. Paquette, "Neural correlates of a mystical experience in Carmelite nuns," *Neuroscience Letters* 405 (2006): 186–90. Be forewarned that, in my view, this paper is written in an unclear fashion. The abstract reads, "The brain activity of Carmelite nuns was measured while they were subjectively in a state of union with God." Nonsense. As the authors reveal in their methods section, the truth is more like "The brain activity of Carmelite nuns was measured *while they attempted to recall a long-past experience* in which they were subjectively in a state of union with God." There is likely to be a very big difference between these two conditions.

21. Beauregard and O'Leary, *The Spiritual Brain*, 276.

22. The paper I discuss is W. T. Harbaugh, U. Mayr, and D. R. Burghart, "Neural responses to taxation and voluntary giving reveal motives for charitable donations," *Science* 316 (2007): 1622–25. It built upon earlier work, showing activation of both the VTA and the nucleus accumbens by anonymous charitable giving: J. Moll, F. Krueger, R. Zahn, M. Pardini, R. de Oliveira-Souza, and J. Grafman, "Human fronto-mesolimbic networks guide decisions about charitable donation," *Proceedings of the National Academy of Sciences of the USA* 103 (2006): 15623–28. Not surprisingly, these experiments have generated all sorts of political discussion. Writing in the *Times Online* (UK), Terence Kealey says, "First, it disproves the Left's belief that only the state will succour the poor. . . . Secondly, of course, it disproves the Right's belief that taxes are unpopular." Oy. http://www.timesonline.co.uk/tol/comment/columnists/guest_contributors/article2204111.ece.

23. I'm not a scholar of philosophy, but it's worth noting that motivations for prosocial behavior have been a topic of intense interest in this field. Kant, for example, wrote that acts driven by feelings of sympathy were not truly altruistic, and were thereby undeserving of praise, because they made the actor feel good. Brain scanning studies of the brain's pleasure circuit suggest that this is a very tough standard to meet.

24. K. Izuma, D. N. Saito, and N. Sadato, "Processing of social and monetary rewards in the human striatum," *Neuron* 58 (2008): 284–94. These authors used a clever control experiment to address the concern that merely seeing positive words was rewarding. They showed the subjects images in which the terms were applied to a fictitious third party, whose image was presented on the video screen. These positive terms failed to activate the reward circuit. There is also a recurring caveat that should be sounded about the interpretation that monetary and social reward activate the same circuit. Of course, these are low-resolution brain imaging studies, and we don't yet know that these two forms of reward activate the exact same neurons. It is formally possible that there are two different circuits in the nucleus accumbens for these two different forms of reward and they are separate when analyzed at the level of individual cells.

25. K. Fleissbach, B. Weber, P. Trautner, T. Dohmen, U. Sunde, C. E. Elger, and A. Falk, "Social comparison affects reward-related brain activity in the human ventral striatum," *Science* 318 (2007): 1305–8.

26. Exodus 20:17, paraphrased a bit.

27. E. S. Bromberg-Martin and O. Hikosaka, "Midbrain dopamine neurons signal preference for advance information about upcoming rewards," *Neuron* 63 (2009): 119–26.

28. Read Montague's theory of how ideas can engage the pleasure/reward circuit in the human brain and thereby cause all kinds of things to happen in human behavior and culture is laid out in his *Why Choose This Book? How We Make Decisions* (New York: Dutton, 2006). Confusingly, this book has been republished in paperback with a different title: *Your Brain Is (Almost) Perfect: How We Make Decisions* (New York: Plume, 2007).

29. While we are imagining that this association is created by strengthening excitatory synapses that drive the VTA, there are other possibilities as well.

Perhaps, for example, there are inhibitory synapses received by VTA dopamine cells that are undergoing LTD. Alternatively, there might be changes in the spike-generating machinery (voltage-sensitive ion channels) of the VTA dopamine neuron that allow it to respond with more spikes to a greenlight-driven synaptic input.

CHAPTER SEVEN: THE FUTURE OF PLEASURE

1. Ray Kurzweil, *The Singularity Is Near* (New York: Viking Penguin, 2005), 163–67.
2. Ray Kurzweil interview in *GOOD* magazine, http://www.good.is/post/going-down-the-rabbit-hole/, posted April 7, 2009.
3. Kurzweil, *The Singularity Is Near,* 198–203.
4. Here's an example from my own lab. We study cellular and molecular processes underlying memory storage in the cerebellum, a part of the brain involved in motor control and motor learning. Together with the labs of our colleagues David Ginty and Paul Worley, we have evidence suggesting that a protein called serum response factor, or SRF, is required for the transition from short-term to long-term memory. If we want to interfere with SRF function to test this hypothesis, we can just look up the sequence of the SRF gene in the mouse genome database and design an RNA probe to block its function. We can then coat the probe on a tiny gold particle and blast it into a mouse neuron using compressed nitrogen gas. Having the mouse genome sequence doesn't give us a "Eureka!" moment just by looking at the SRF gene sequence, but it is a very useful tool.
5. Kudos to philanthropists Paul G. Allen and Jody Allen Patton, who have provided the initial funding for this project at the Allen Institute for Brain Science. You can see the images at http://www.brain-map.org.
6. M. J. Frank, B. B. Doll, J. Oas-Terpstra, and F. Moreno, "Prefrontal and striatal dopaminergic genes predict individual differences in exploration and exploitation," *Nature Neuroscience* 12 (2009): 1062–68; J.-C. Dreher, P. Kohn, B. Kolachana, D. Weinberger, and K. F. Berman, "Variation in dopamine genes influences responsivity of the human reward system," *Proceedings of the National Academy of Sciences of the USA* 106 (2009): 617–22.
7. J. V. Becker and V. L. Quinsey, "Assessing suspected child molesters," *Child Abuse & Neglect* 17 (1993): 169–74.
8. In this study, the pedophile and nonpedophile populations were matched for age and socioeconomic status: B. Schiffer, T. Krueger, T. Paul, A. de Greiff, M. Forsting, N. Leygraf, M. Schedlowski, and E. Gizewski, "Brain response to visual stimuli in homosexual pedophiles," *Journal of Psychiatry and Neuroscience* 33 (2008): 23–33.
9. The effects of drinking alcohol while on Antabuse are similar to the intense flushing reaction experienced by those who harbor a deletion of one or both copies of their alcohol dehydrogenase type 2 gene. This deletion is particularly common among East Asians and Native Americans.
10. F. M. Orson, B. M. Kinsey, R.A.K. Singh, Y. Wu, T. Gardner, and T. R. Kosten, "Substance abuse vaccines," *Annals of the New York Academy of Sciences* 1141 (2008): 257–69.

11. The neurotransmitter acetylcholine works by activating two broad classes of receptor on neurons: slow-acting muscarinic receptors and fast-acting nicotinic receptors. Nicotine, of course, activates the latter. Nicotine receptors are assembled in a combinatorial manner from different subunits. Varenicline binds to one specific type of nicotinic receptor complex, called $\alpha 4 \beta 2$. It's neither a pure activator nor a pure blocker of this receptor. Rather, it functions as a "partial agonist," which means that it binds the receptor but activates it weakly, thereby reducing its activation by acetylcholine or nicotine.

12. For an overview of the near-term prospects for new anti-addiction drugs see G. F. Koob, G. K. Lloyd, and B. J. Mason, "Development of pharmaco-therapies for drug addiction: a Rosetta Stone approach," *Nature Reviews Drug Discovery* 8 (2009): 500–515.

13. Glutamate receptors are divided into two large families: the fast-acting ionotropic receptors, which open ion channels to produce rapid electrical signaling (these carry the bulk of information flow in your brain); and the slow-acting metabotropic receptors, which activate or inhibit ion channels or enzymes through an intermediate form of signaling called a G-protein. Metabotropic glutamate receptors generally have actions that last for seconds to tens of seconds, and they are often the trigger for other plastic processes that last much longer still. There are eight major forms of metabotropic receptor in the human brain with different distribution patterns and different electrical and biochemical sequelae.

14. C. Chiamulera, M. P. Epping-Jordan, A. Zocchi, C. Marcon, C. Cottiny, S. Tacconi, M. Corsi, F. Orzi, and F. Conquet, "Reinforcing and locomotor stimulant effects of cocaine are absent in mGluR5 null mutant mice," *Nature Neuroscience* 4 (2001): 873–74.

15. While, at the time of this writing, mGluR5 antagonists are not being evaluated as anti-addiction therapies in clinical trials, these drugs are being assessed for a number of other uses, including reduction of anxiety, pain relief, and control of seizures.

16. E. K. Miller and M. A. Wilson, "All my circuits: using multiple electrodes to understand functioning neural networks," *Neuron* 60 (2008): 483–88.

17. M. A. Nicolelis and M. A. Lebedev, "Principles of neural ensemble physiology underlying the operation of brain-machine interfaces," *Nature Reviews Neuroscience* 10 (2009): 530–40; N. G. Hatsopoulos and J. P. Donoghue, "The science of neural interface systems," *Annual Review of Neuroscience* 32 (2009): 249–66.

18. It turns out that there are a number of ways of introducing fluorescent molecules into neurons in the living mouse brain. One way is to use genetic tricks to make strains of mice that have sequences spliced into their DNA to command the neurons to make fluorescent proteins. Another way is to introduce this DNA by injecting an engineered virus into the brain. The virus infects certain neurons and causes them to produce the fluorescent proteins. Yet another way is to use fluorescent molecules that are not proteins and inject these small molecules directly into the brain where neurons (and sometimes glial cells) can take them up. Two reviews that discuss the use of in vivo multiphoton imaging are O. Garaschuk, R. I. Milos, C. Grienberger,

N. Marandi, H. Adelsberger, and A. Konnerth, "Optical monitoring of brain function in vivo: from neurons to networks," *Pflügers Archiv* 453 (2006): 385–96; A. Holtmaat, T. Bonhoeffer, D. K. Chow, J. Chuckowree, V. De Paola, S. B. Hofer, M. Hübener, T. Keck, G. Knott, W. C. Lee, R. Mostany, T. D. Mrsic-Flogel, E. Nedivi, C. Portera-Cailliau, K. Svoboda, J. T. Trachtenberg, and L. Wilbrecht, "Long-term, high-resolution imaging in the mouse neocortex through a cranial window," *Nature Protocols* 4 (2009): 1128–44.

19. V. Gradinaru, K. R. Thompson, F. Zhang, M. Mogri, K. Kay, M. B. Schneider, and K. Deisseroth, "Targeting and readout strategies for fast optical neural control in vitro and in vivo," *Journal of Neuroscience* 27 (2007): 14231–38.

20. This paper is really interesting: D. Nutt, L. A. King, W. Saulsbury, and C. Blakemore, "Development of a rational scale to assess the harm of drugs of potential misuse," *Lancet* 369 (2007): 1047–53. It attempts to measure the social and individual harm produced by a number of psychoactive drugs, from alcohol and nicotine to ecstasy and heroin, and then compares that ranking with the penalties for use of those substances in the United Kingdom. The conclusion: In many cases, the penalties do not match up well with the harm.

INDEX